W9-AEZ-803

THE HUMAN MACHINE

THE HUMAN MACHINE

by R. McNeill Alexander
Professor of Zoology in the University of Leeds, England

Illustrated by Mark Iley & Sally Alexander

COLUMBIA UNIVERSITY PRESS

Planned and produced by
Natural History Museum Publications
Cromwell Road, London SW7 5BD

Published in the United States of America
by Columbia University Press
New York Oxford

Designed by Michael Morey

Library of Congress Cataloging-in-Publication Data
Alexander, R. McNeill
The human machine / by R. McNeill Alexander;
illustrated by Mark Iley & Sally Alexander.
p. cm.
Includes index.
ISBN 0–231–08066–2
1. Human mechanics. I. Title.
QP303.A58 1992
612.7′6—dc20
92–10517
CIP

ISBN 0–231–08066–2

c 10 9 8 7 6 5 4 3 2 1

Typeset by J&L Composition Ltd, Filey, North Yorkshire, England
Printed by Cambus Litho, Nerston, East Kilbride, Scotland

CONTENTS

PREFACE

This book views the human body and its movements as an engineer might view a structure or a machine. It discusses all kinds of human movement, from weight-lifting to speech, explaining the mechanical principles on which they all depend. There are accounts in it of everyday activities such as walking and breathing; of sporting activities including many aspects of athletics; and of some of the medical problems that arise when the body's mechanical systems fail.

It is not a book for specialists in biomechanics, but for everyone who is interested in the working of their own bodies. I have assumed very little prior knowledge of mechanics or indeed of any branch of science, and have tried to explain everything as simply as possible.

Many people apart from myself have played important parts in the production of this book. Myra Givans and Isobel Smith supervised its publication, and Michael Morey is responsible for the design. Mark Iley made the anatomical drawings and Sally Alexander the technical ones. Alison Long appears in most of the photographs, that were taken specifically for the book by Neil Fletcher. Dr Celia Scully and many others helped me with information and illustrations. I am very grateful to them all.

R. McNeill Alexander

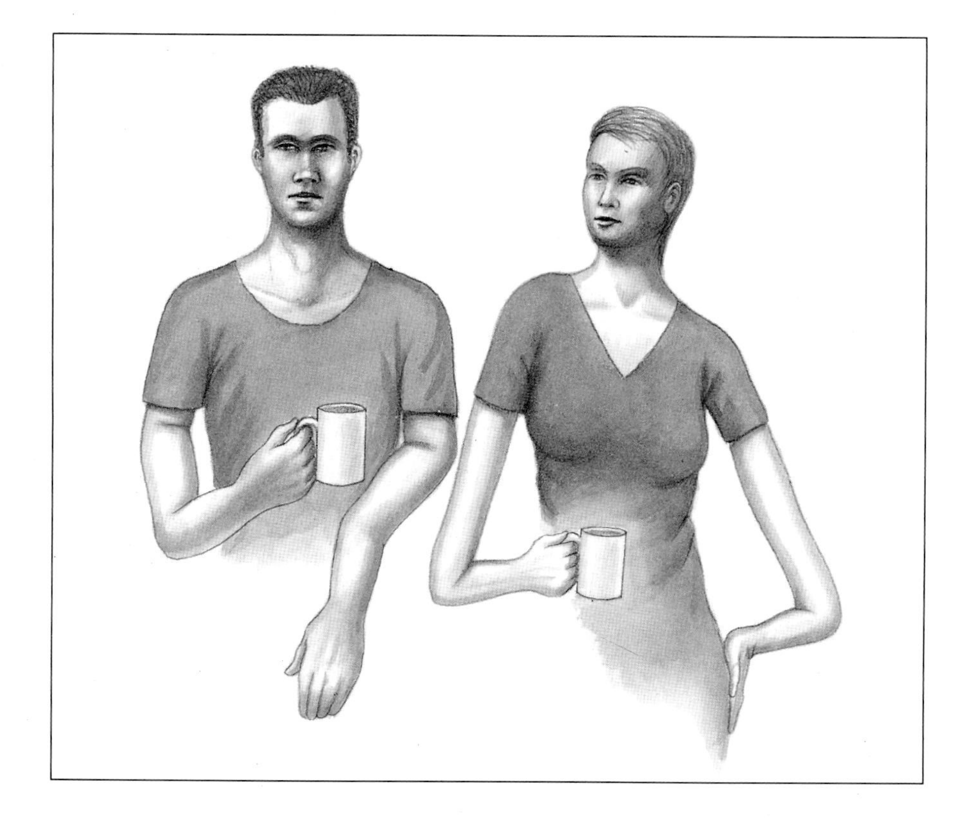

FIG. 1.1 *Would our arms be more useful if they had extra joints or were differently proportioned?*

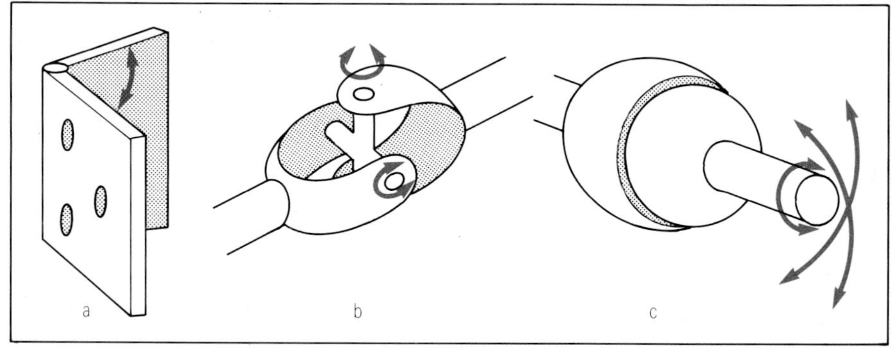

FIG. 1.2 *A hinge joint (a) allows one degree of freedom of movement, a universal joint (b) allows two, and a ball-and-socket joint (c) allows three.*

CHAPTER ONE

REACHING

Our skeletons are made of rigid bone so we need joints to let us move, but what kinds of joint do we need, and how many? Would our bodies work even better if we had two elbows in each arm, or if we had longer upper arms and shorter forearms (FIG. 1.1)? This chapter tries to answer such questions.

It starts by describing basic kinds of joint and showing how they are represented in the human arm. A hinge is one of the simplest kinds. The only movement it allows is rotation about a single axis: you can open and close a hinged door but you cannot move it in any other way. To describe the position of a door you need only state one number, for example the angle of the door to the wall or the distance from its edge to the jamb. In the technical terms of engineering, a hinge allows only one degree of freedom of movement. Your elbow is a hinge.

Universal joints, made from two hinges with their axes at right angles, allow two degrees of freedom (FIG. 1.2). Such joints are used in the drive shafts of cars, which have to transmit torque from the engine to the wheels while allowing the wheels to bounce up and down on their springs. Whilst the car is stationary, one hinge in its shaft (with its axis horizontal) would be enough to let the wheels move up and down, but when the car moves the shaft rotates and two hinges are needed. Between them, they allow the wheels to move up and down at any stage of the shaft's rotation. To describe the position of a universal joint you have to give two numbers, for example the angles at the two hinges. Your wrist is a universal joint. It allows you to make the two kinds of movement shown in FIGS. 1.3a,b (or any combination of these kinds) but prevents other movements. In particular, it does not allow the movement suggested in FIG. 1.3c: if you hold the bones of your forearm stationary (grip them firmly with your other

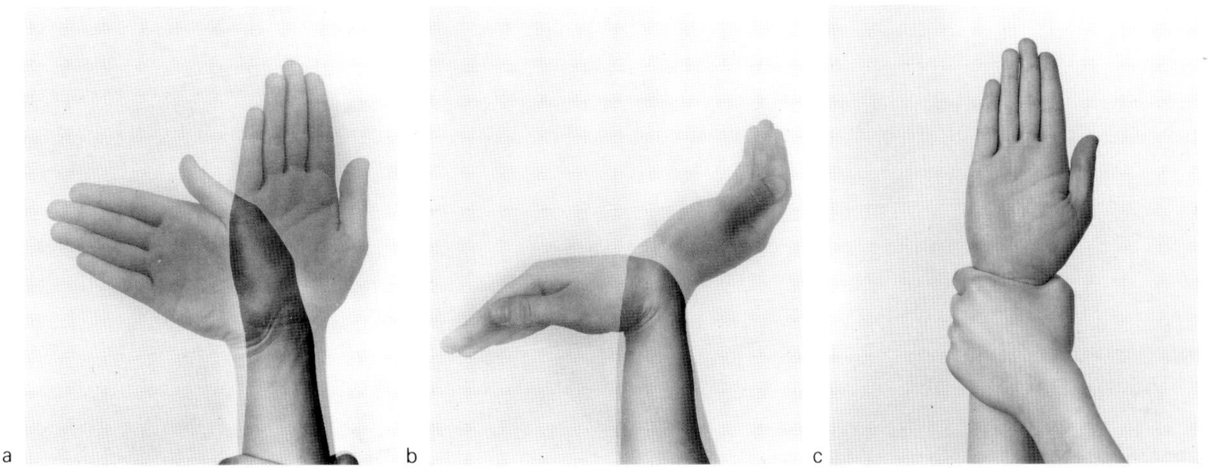

FIG. 1.3 *The wrist allows two degrees of freedom of movement (a and b) but not a third (c).*

hand) you cannot turn your hand from palm-up to palm-down.

A ball-and-socket joint consists of a ball in a tight-fitting spherical socket and allows three degrees of freedom of movement. It allows rotation about *any* axis that passes through the centre of the ball but we nevertheless say it allows just three degrees of freedom because three numbers are enough to describe its position. For example, if the socket is kept stationary, the position of the movable arm can be described by giving its compass direction, the angle at which it is tilted relative to the horizontal, and its angle of rotation about its own long axis. Your shoulder has a ball-and-socket joint. To see what it can do, start with both arms by your sides with elbows bent, like the right arm of the woman in FIG. 1.4. Keeping other joints fixed you can rotate the arm about a vertical axis through the shoulder, making your hand move out to the side (FIG. 1.4a, left arm), you can rotate it about a horizontal front-to-back axis (b) or you can rotate it about a horizontal side-to-side axis (c). You can also rotate it about any other axis through the centre of the shoulder, but all possible movements can be described as combinations of these three.

We have one bone (the humerus) in the upper arm and two (radius and ulna) in the forearm (FIG. 1.5). The ball at the top of the humerus fits into a socket in the scapula (shoulder blade). In man-made ball-and-socket joints the socket is rather more than half a sphere, so that the ball cannot be pulled out of it. The socket in the shoulder blade is much less than half a sphere, and the joint is held together only by ligaments and muscles.

The ligaments of the joints in our arms and legs are strong, flexible bands of tissue, white in colour, that consist mainly of a protein called collagen. They are flexible because the collagen is divided into a large number of parallel fibres (a wire rope is more flexible than a solid rod of the same thickness, made

a

b

FIG. 1.4 *The shoulder joint allows three degrees of freedom of movement.*

of the same steel). Most joints are held together mainly by ligaments, but the shoulder depends mainly on muscles. It is very difficult to think how ligaments could be arranged at a ball and socket joint, so as to allow a wide range of movements and yet always hold the bones tightly together. Muscles, however, can be kept tight in all positions of the joint, lengthening to allow the joint to move one way and shortening to take up the slack as it moves the other. The muscles that do this job lie on and under the scapula. They attach at one end to the flat surfaces of the scapula and at the other to the head of the humerus, all around the joint surface.

Because it depends on muscles to hold it together the shoulder is rather easily dislocated (the ball gets pulled out of the socket). This often happens in ice hockey accidents, and in other falls. A very strong push on the front of the left arm in FIG. 1.4b, would be apt to make the head of the humerus spring out from the front of its socket. Fortunately, dislocation

c

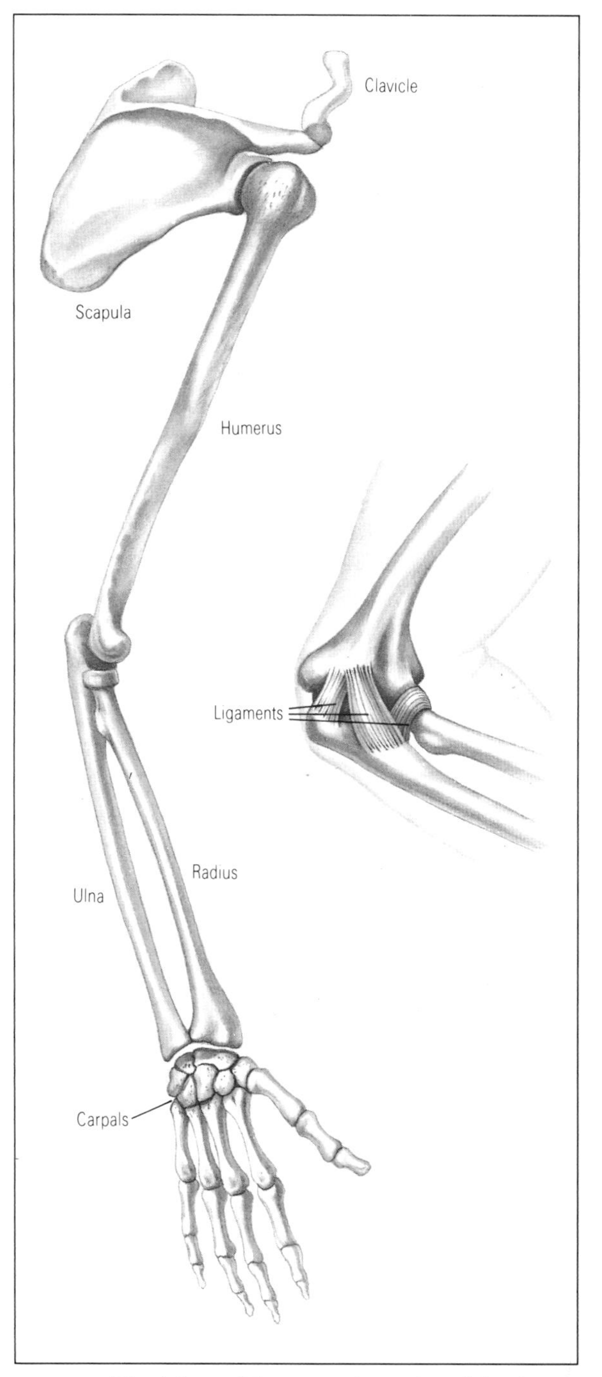

FIG. 1.5 *The skeleton of the arm, with an enlarged detail.*

does not generally cause much damage to the surrounding tissues and it is not too difficult to pull the humerus back into place.

There is a hinge joint at the elbow, between the humerus and the ulna. A spool-shaped surface on the end of the humerus fits into a close-fitting hollow in the ulna, so while the bones are held firmly together the only possible movement is rotation about the axis of the spool. This joint has ligaments on either side, which attach to the humerus close to the axis of the joint. They allow it to move freely but hold it firmly together, avoiding the problem suggested in FIG. 1.6. Obviously, the fibres of a ligament cannot *all* be attached precisely at the axis of a joint, but ligaments can stretch a little so this may not matter.

The radius and ulna are jointed together in a way that allows us to make the movement needed to use a screwdriver (FIG. 1.7). At the elbow, the cylindrical end of the radius fits into a groove in the ulna. It is held in place by a ligament, attached to the ulna at both ends, that loops around it. The shafts of the two bones are connected by a strong membrane of collagen fibres. At the wrist end the ulna fits into a groove in the head of the radius where it is held in place by the ligaments connecting both bones to the bones of the wrist and to a fibrous disc in the wrist joint. Thus the radius can rotate about an axis that runs obliquely between it and the ulna. The practical effect (allowing the hand to make screwdriver movements) is the same as if the elbow were a simple hinge and the wrist a ball-and-socket instead of a universal joint.

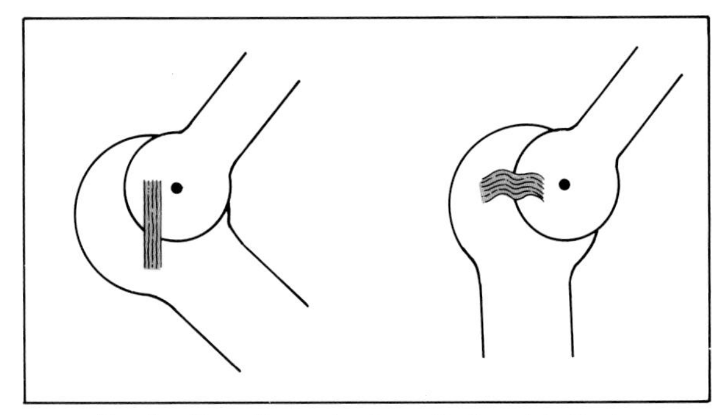

FIG. 1.6 *An elbow ligament whose attachment to the humerus was off the axis of the joint would become tight at one position of the joint (stopping further movement) and slack at another (allowing dislocation).*

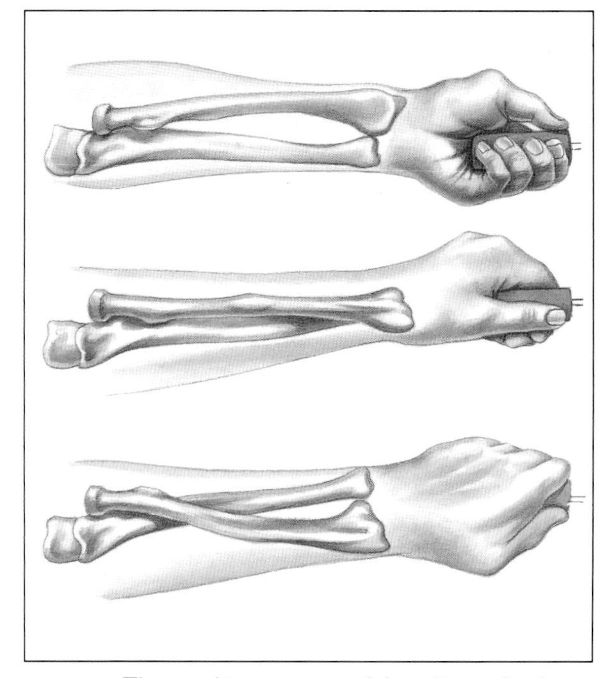

FIG. 1.7 *The screwdriver movement of the radius on the ulna.*

The wrist has eight small bones (the carpal bones) held together by a complicated arrangement of ligaments whose functions are not fully understood.

Thus the shoulder is a ball-and-socket joint allowing three degrees of freedom of movement; the joints between humerus and ulna and between ulna and radius are hinges each allowing one degree of freedom; and the wrist is a universal joint allowing two degrees of freedom. Ignoring movements within the hand (which will be discussed in Chapter 2) and the movements of the shoulder blade (as when you shrug your shoulders), the arm as a whole has seven degrees of freedom of movement. Is this the best number? Before trying to answer the question we must go back to basic principles.

If I mark a dot on a piece of graph paper I can describe its position by giving just two measurements, for example its x and y coordinates (FIG. 1.8a). In other words, a point moving freely in a

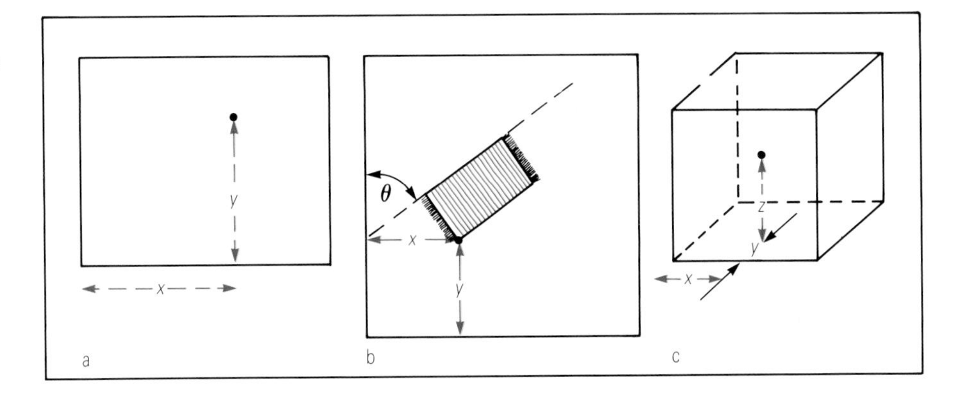

FIG. 1.8 *How the positions of points and of a rug can be described.*

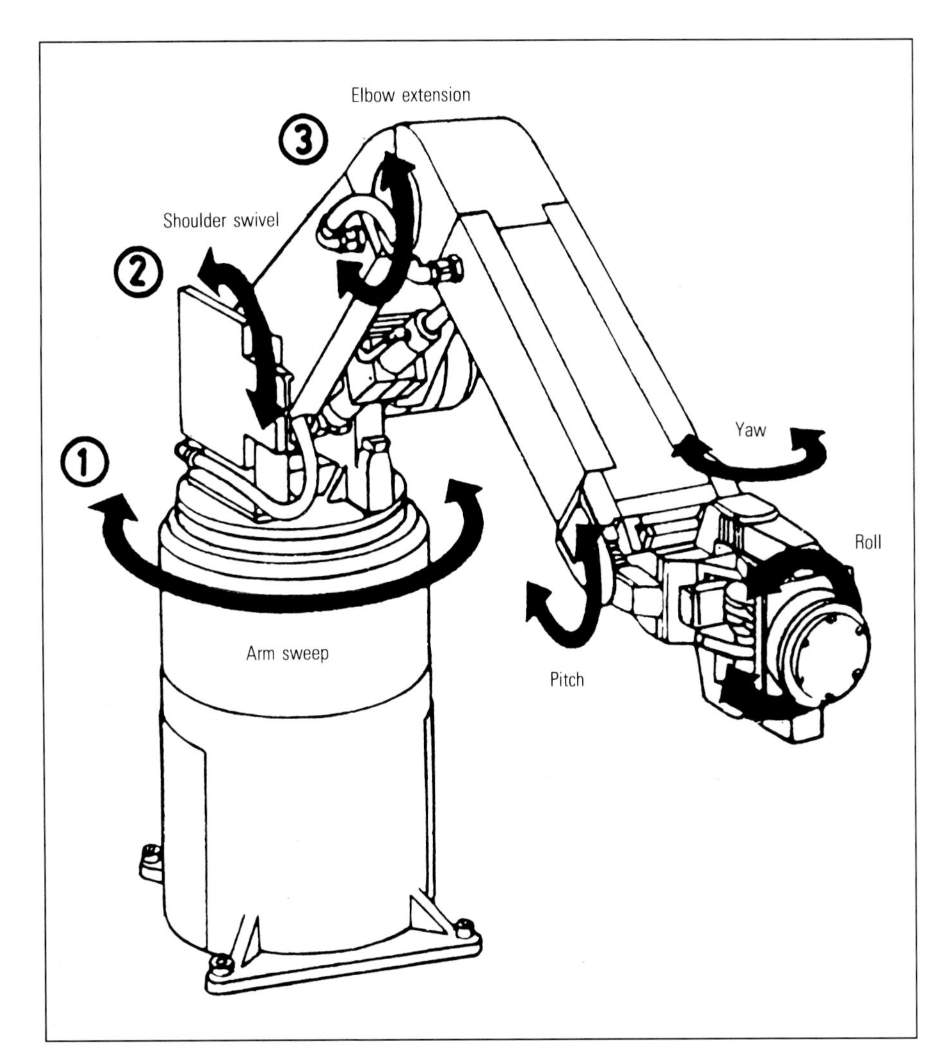

FIG. 1.9 *An industrial robot. This is the Cincinnati Milacron T3. From F. N.-Nagy and A. Siegler (1987)* Engineering Foundations of Robotics *Prentice-Hall, Englewood Cliffs, N.J.*

plane has two degrees of freedom of movement. However, to describe the position of a rug on the floor three measurements are needed, for example, the distances x and y and the angle θ (FIG. 1.8b). Now think of a point moving in the space of a room: three quantities (such as x, y and z, FIG. 1.8c) will describe its position. Finally, think of a rigid body moving in space. To describe its position six measurements are needed, for example, three distances to tell where one particular point on the body is and three angles to describe how it is tilted (in any possible direction) relative to its surroundings. A rigid body moving freely in space has six degrees of freedom of movement. Ignore for the moment the movements that your hand can make and think of it as a rigid body. If you want to be able to put it into any position (within arm's reach) and tilt it to any angle you need an arm whose joints allow six degrees of freedom of movement. Your arm joints actually allow seven, which seems one more than necessary (or even more, if we take account of the mobility of the scapula).

We will look for a possible explanation by asking how engineers build arm-like devices. Robots are used in factories for assembling cars, and many other tasks. FIG. 1.9 shows an example. It can move a tool such as a drill (attached at the end of the arm) to any position within arm's reach, and hold it at

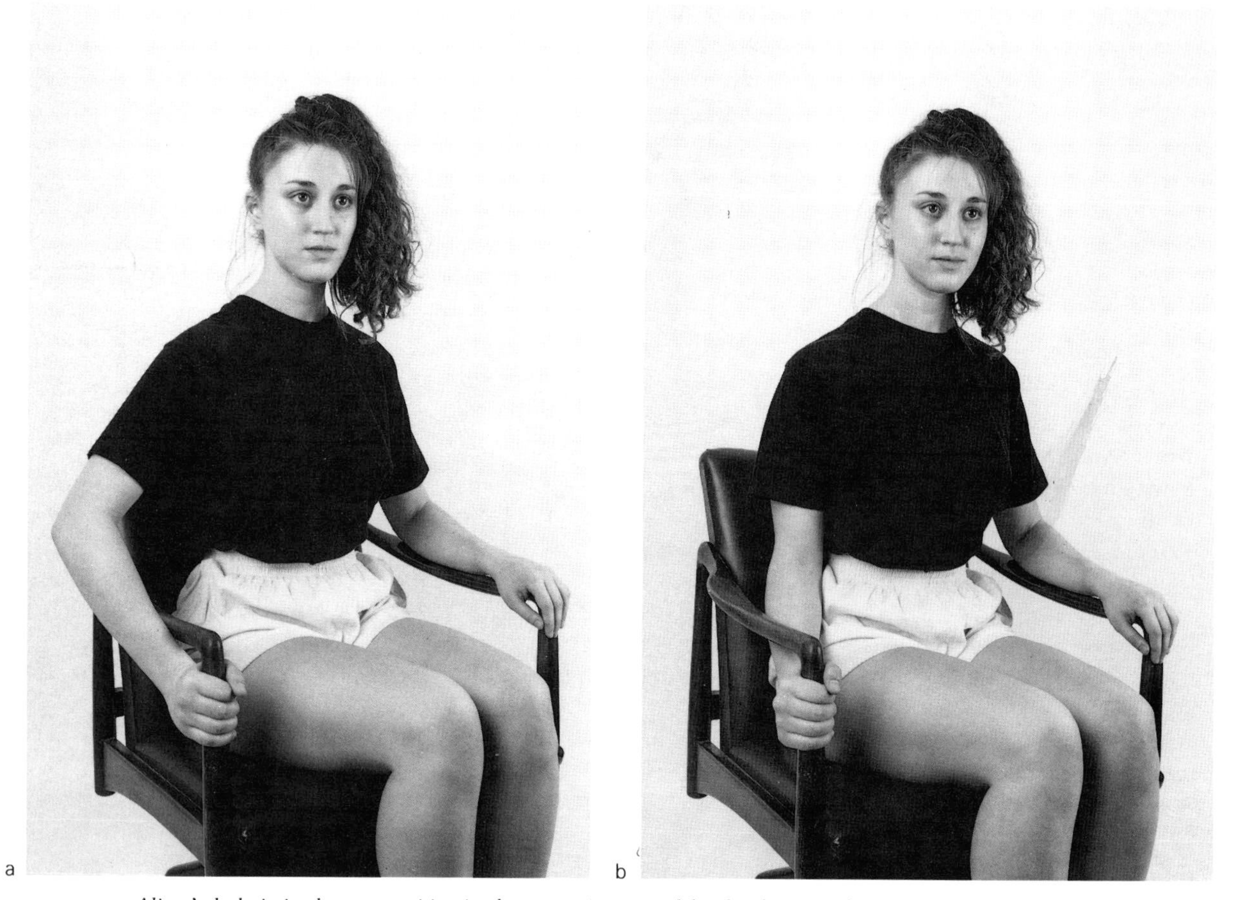

FIG. 1.10 *Alison's body is in the same position in these two pictures and her hand grasps the same part of the chair, in the same way, but in one case she is reaching over and in the other case under the arm.*

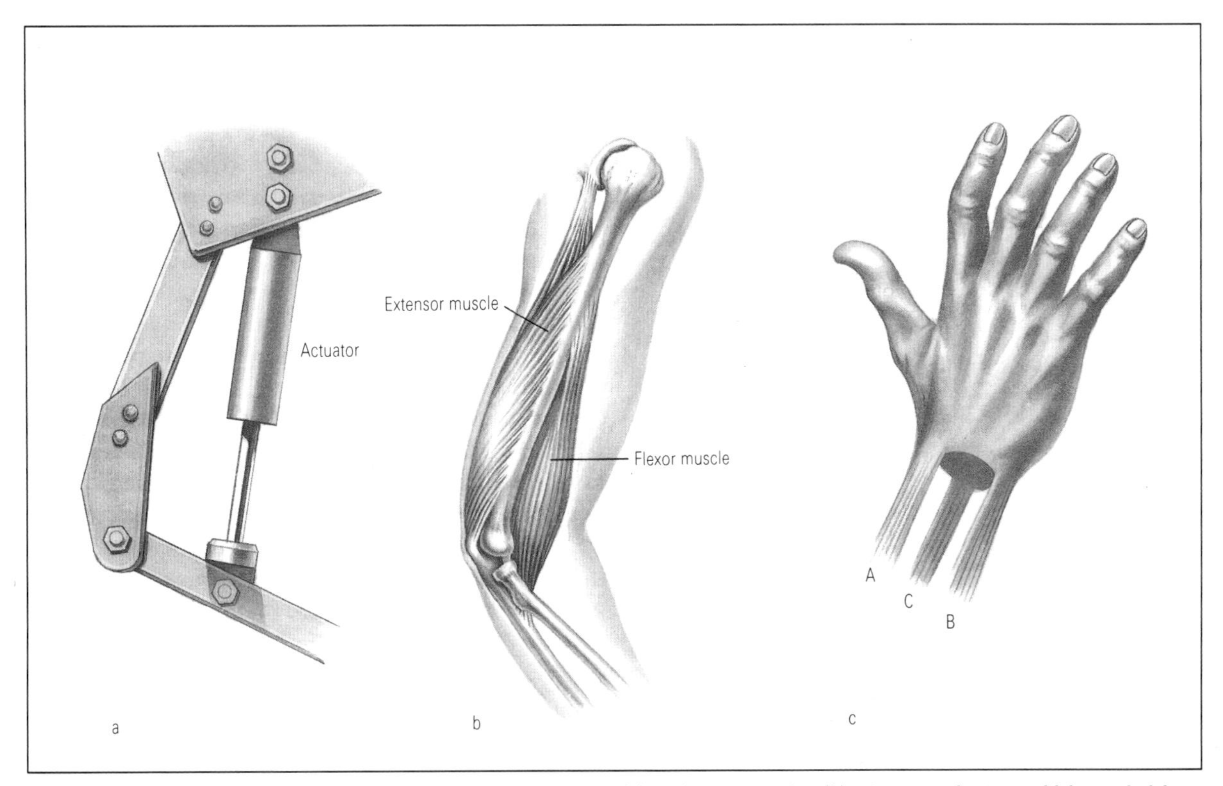

FIG. 1.11 *A hinge joint can be worked by one hydraulic cylinder (a) or by two muscles (b). A universal joint could be worked by three muscles (c).*

any angle. Its joints allow the necessary six degrees of freedom (two at the base, one at the 'elbow' and three at the 'wrist'). Many other industrial robots likewise have just six degrees of freedom, though the details of the design vary. For example, some have the base running on rails and no 'arm sweep' hinge: the possibility of moving along the rails (one degree of freedom) replaces that of turning about one hinge. However, some industrial robots have seven degrees of freedom in their arms. One example resembles FIG. 1.9 but has an extra joint between the base and the elbow.

An arm with six degrees of freedom allows no choice. There is only one combination of joint angles that will place the hand in any particular position, and the arm must take a particular route from shoulder to wrist. Such an arm is easily prevented by obstructions from reaching where it is wanted: it cannot reach round things that are in the way. An extra degree of freedom allows more flexibility. For example, the woman in FIG. 1.10 can get her hand to the same position in two different ways. She would not have this choice if her arm had only six degrees of freedom.

The joints of robots are commonly worked by hydraulic cylinders which push to extend a joint or pull to bend it. One hydraulic cylinder is enough to work a hinge joint (FIG. 1.11a), and an arm needs as many cylinders as it has degrees of freedom. Our joints are worked by muscles that can pull but cannot push, so two muscles are needed to work a hinge joint. The one that bends the joint is called the flexor, the one that straightens it the extensor (FIG. 1.11b).

That may suggest that an arm with seven degrees of freedom needs fourteen muscles, but it could actually be worked by fewer. For example, a wrist

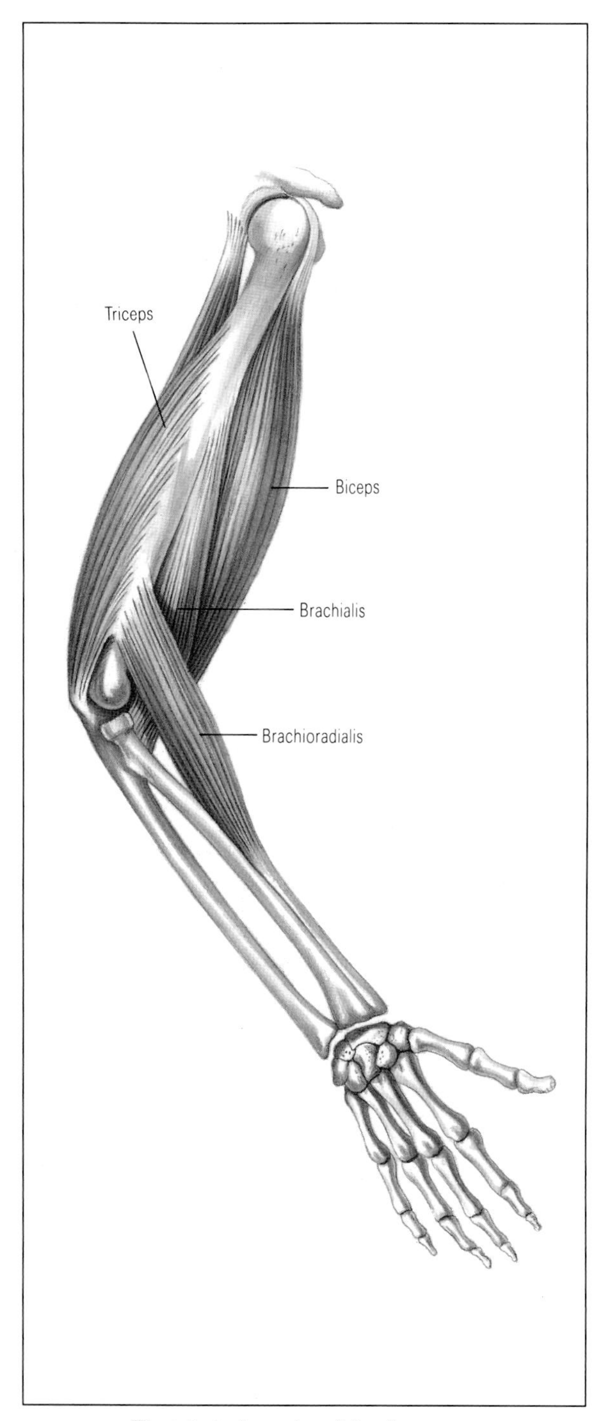

FIG. 1.12 *The principal muscles of the elbow.*

with two degrees of freedom could be worked by just three muscles. In FIG. 1.11C muscle A tends to extend the wrist and to pull it towards the thumb side; muscle B to extend it and pull it to the little finger side; and muscle C to flex it. Any other movement of the wrist could be executed by two of these muscles acting together. For example, A and B can extend the wrist without any movement in the plane of the fingers. The same argument can be used (with some ingenuity) to show how an arm with seven degrees of freedom could be worked by eight muscles.

Our arms actually have far more muscles than that. Ignoring the muscles that connect the scapula to the ribs and the many muscles in the forearm and hand that serve to work our fingers and thumb (which will be described in Chapter 2), the arm still has 22 distinct, separately-named muscles. The wrist has five muscles to work its two degrees of freedom: two side by side in the position of A (FIG. 1.11C), one in the position of B and two on the palm side of the joint. There are three muscles whose principal function is to power the screwdriver movements of radius on ulna (one degree of freedom). Five more muscles cross the elbow joint, working the one degree of freedom of the hinge between humerus and ulna, and nine more muscles work the shoulder joint.

FIG. 1.12 shows the principal muscles of the elbow. The triceps has three parts (its name means 'three heads'), one starting on the scapula and two on the back of the humerus. These join up to attach to the end of the ulna and serve to extend the joint. The other three muscles in the illustration are flexors. The biceps ('two heads') runs from the scapula to the radius, the brachioradialis from the humerus to the radius and the brachialis from the humerus to the ulna. Many muscles cross two or more joints and affect them all. For example, the biceps tends to move the shoulder as in FIG. 1.4C, to bend the elbow and to rotate the radius about the ulna, turning the palm of the hand upward. Which movement occurs when it shortens depends on what the other muscles are doing.

Your upper arm (from shoulder to elbow) is about

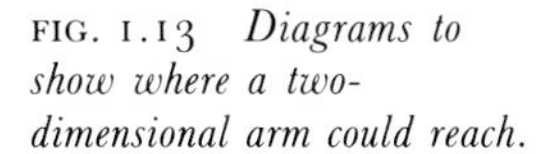

FIG. 1.13 *Diagrams to show where a two-dimensional arm could reach.*

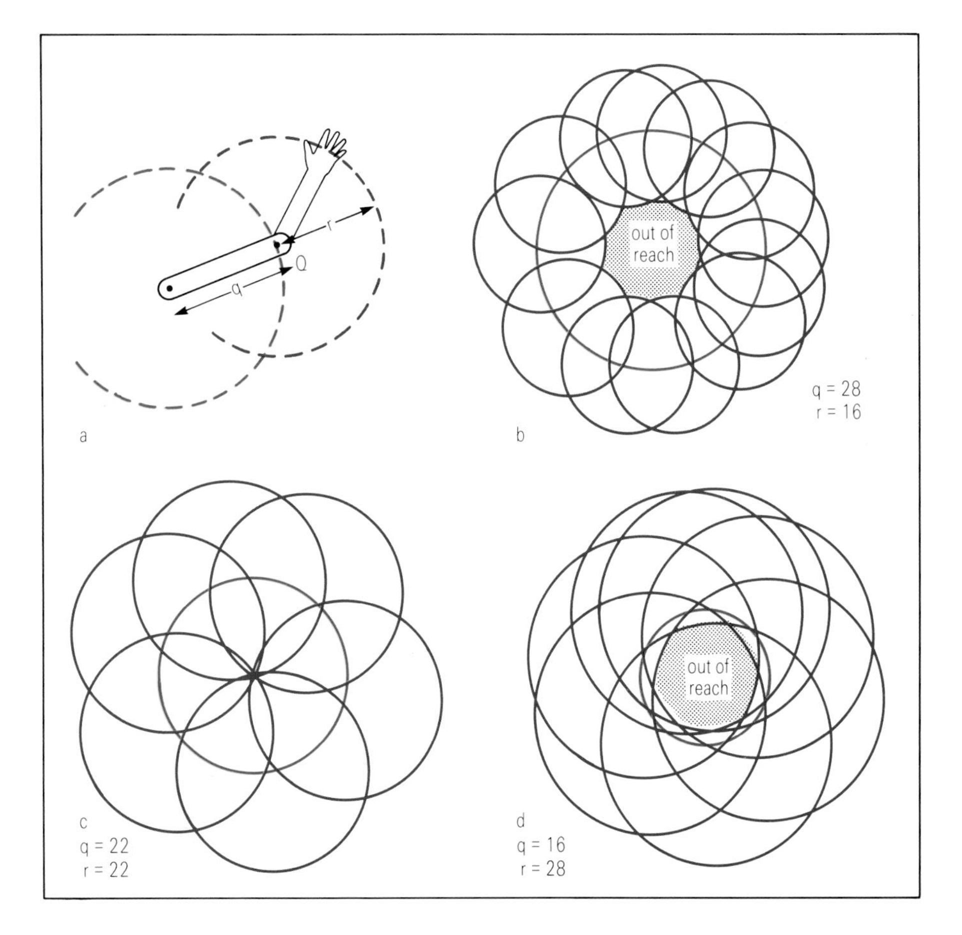

the same length as your forearm (from elbow to wrist). Similarly, the robot arm of FIG. 1.9 has two main segments of about equal length. Is this coincidental? To see why it may not be, imagine the woman of FIG. 1.1 trying to touch her right armpit with her right hand.

We can make the principle clearer by imagining a two-dimensional robot arm consisting of two segments, of lengths q and r (FIG. 1.13). The shoulder and elbow are hinge joints with axes perpendicular to the paper. When the shoulder moves, the elbow moves around it in a circle of radius q. When the shoulder is locked to keep the elbow at Q, the hand can be moved around Q in a circle of radius r. We can find all the points that the hand can reach by drawing a circle of radius q and then lots of circles of radius r, centred at points all around the first circle. The diagram (FIG. 1.13b–d) shows patterns of this kind for three arms of the same total length ($q + r$). All can reach to the same maximum distance from the shoulder, but only the one that has q and r equal can reach to the centre.

Our joints cannot make complete revolutions as implied in FIG. 1.13, and our arms move in three dimensions, not two. For these reasons, the geometry of reaching is much more complicated than the diagram suggests. A much more elaborate diagram would be needed to explain why there is a spot between my shoulder blades that I cannot scratch.

CHAPTER TWO

HANDLING

This chapter is about the movements our hands can make and about how we use them for grasping and manipulating things. We will start by looking for the simplest possible way of holding things firmly. A crab's claw seems painfully well adapted for gripping but it is also rather simple, so let us think about what it can do. Imagine a crab holding a metal bar. FIG. 2.1 has three axes drawn at right angles to each other to define the directions of movement that we will consider. The Y axis is also the axis of the bar, and the Z axis passes through the points of contact of the upper and lower jaws with the bar. We will think about all the ways in which we could try to move the bar and work out whether it is rigidly held. The bar cannot move up and down along the Z axis, because the jaws are in the way. Unless it is remarkably slippery, friction will prevent it from moving forward and back along the X axis or from side to side along the Y axis. Friction will also prevent it from rotating about the X axis or the Y axis, but there is nothing to prevent it from rotating about the Z axis, the line through the two points of contact. The crab's grip has removed all but one of the bar's six degrees of freedom of movement.

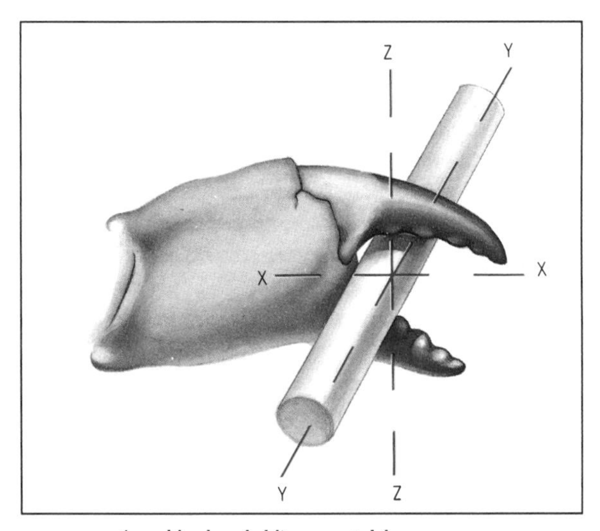

FIG. 2.1 *A crab's claw holding a metal bar.*

The situation would be different if the bar were not metal, but made of something softer like rubber. The jaws of the claw would bite into it a little so that each made contact along a line, not merely at a point, and rotation about the Z axis would be prevented.

Of course, no object is infinitely hard. A crab with jaws that could grip forcefully enough could hold even a hard metal bar quite firmly. However, can you imagine a crab holding an egg without breaking it?

One solution to the problem of holding hard things is to have soft finger tips. You can hold a book reasonably firmly between finger and thumb as in FIG. 2.2a, because the pressure deforms the flesh, giving an area of contact instead of a point. Rotation of the book involves sliding at all parts of the areas of contact that are not precisely on the axis of rotation, and this sliding is resisted by friction. However, because the area is quite small, no part of it is far from its centre, and it is quite difficult to prevent rotation about an axis through the centre of the area. You must squeeze a book very tightly between finger and thumb to hold it as in FIG. 2.2a. It is much easier to hold it with a thumb and two fingers, as FIG. 2.2b.

A hard object can be held even by hard fingers, if there are three points of contact that are not all in line with each other. This suggests that the minimum satisfactory number of fingers is three (or two fingers and a thumb).

a

FIG. 2.2 *How big a book can you hold like this (a) between one finger and thumb and like this (b) using an additional finger?*

b

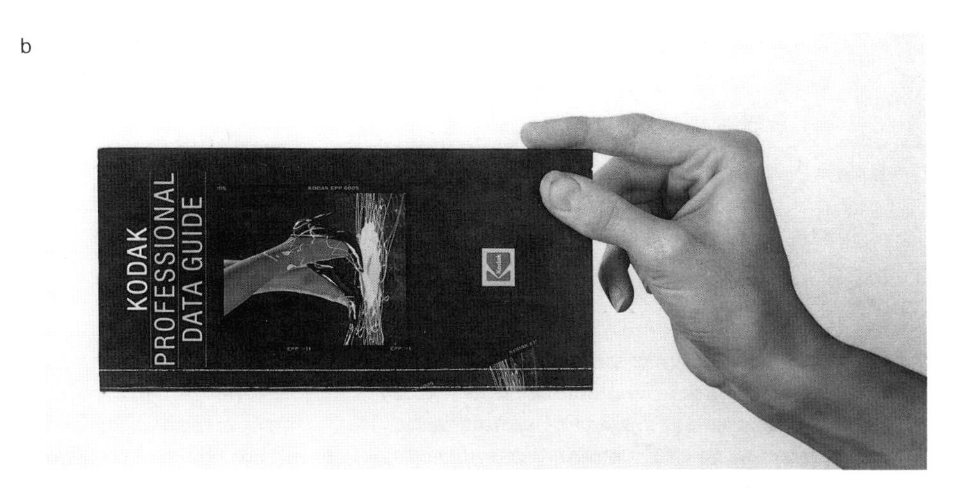

FIG. 2.3 *Knotting a rope.*

There is no apparent need for more than three fingers to grip any single rigid object, but some of the tasks we perform with our hands involve holding several things at once, or several parts of a flexible object. The hands knotting a rope in FIG. 2.3 are each holding it in three places, with different fingers. Having five fingers rather than three enables us to do complicated things like this. If we had more than five fingers we might be able to do even more complicated tasks.

We have already mentioned the carpal bones, the eight small bones of the wrist. The other bones of the hand are the five metacarpals (in the palm) and fourteen phalanges (two in the thumb and three in each finger). They are connected by a very large number of movable joints (FIG. 2.5). Those between the phalanges of each finger are hinges, allowing just one degree of freedom of movement each. The joints at the bases of the fingers, between the metacarpal

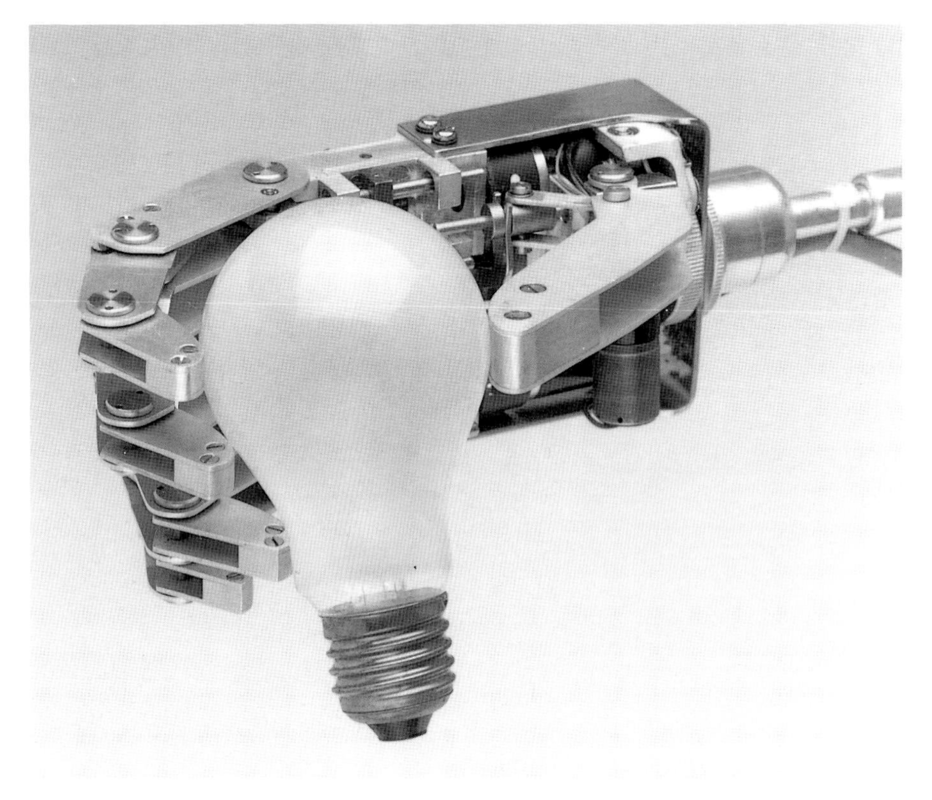

FIG. 2.4 *The Southampton prosthetic hand. Photograph supplied by Dr P.J. Kyberd, Department of Electrical Engineering, University of Southampton.*

and the first phalanx, are universal joints allowing two degrees of freedom. You use one of these degrees of freedom when you flex your fingers to make a fist and the other when you spread your fingers like a fan. The thumb is differently constructed from the fingers, with the universal joint between carpal and metacarpal, not between metacarpal and first phalanx. This enables us to 'oppose' it to the other fingers so that we can grip things effectively between the thumb and one or more fingers. The metacarpals of the index and middle fingers can hardly be moved at all relative to the carpals but those of the ring and little fingers move a little, when we arch our palms. That is why joints are shown at the bases of some of the metacarpals in FIG. 2.5b but not at the bases of others. This diagram shows twelve hinge joints and five universal joints within the hand, allowing a total of 22 degrees of freedom of movement. In contrast, the artificial hand shown in FIG. 2.4 has only four separately controllable degrees of freedom: two in the thumb, one in the index finger and one shared by the other fingers (which bend together).

To put a finger tip in a desired position relative to your palm you need finger joints allowing three degrees of freedom of movement (assuming that you only want to get a point on the tip to the right place and are not concerned about the orientation of the finger). Three hinge joints whose axes were not all parallel to each other would do, or one hinge and one universal joint. Our fingers actually have four degrees of freedom each, provided by two hinges and a universal joint. That makes us better able to curl our fingers round things that we are gripping. To put this in a different way, the extra degree of freedom allows a range of finger positions, for given positions of the palm and finger tip. Similarly, an extra degree of freedom in the arm allows a range of arm positions, for given positions of shoulder and hand (FIG. 1.10).

Anatomy books describe 29 muscles that work the hand, but some of these have several distinct parts, connected to the bones by separate tendons, that seem able to work separately. If we count these parts as distinct muscles the effective number of muscles becomes 38. This is almost twice the number of degrees of freedom, which should not seem surprising after the discussion of arm muscles in Chapter 1. However, the arrangement of muscles is complicated.

There is not simply a pair of muscles to work each degree of freedom. Instead, most of the hand muscles cross several joints and affect the movements of them all.

Some of these muscles are in the hand itself but many are in the forearm, connected to the bones of the hand by long tendons. (Tendons attach muscles to bones whereas ligaments, made of similar material, connect bones to each other.) My colleague Dr Alison Cutts and I dissected all the hand muscles from an arm that had been severed in an accident. There were 20 muscles in the hand itself, with a total mass of only 91g. The forearm contained another 20 muscles with a total mass of 766g; yet of these, nine muscles with a mass of 401g, were muscles that serve to move the fingers and thumb. The mass of the intact hand must have been about 600g. If all the hand muscles had been in the hand itself it would have had to be much bulkier, and correspondingly unwieldy.

You can feel the tendons of some of these muscles in your own hand. When you hold your hand as in FIG. 2.6, a hollow appears at the base of the thumb, its side walls formed by the tendons of the two extensor muscles of the thumb. This hollow is called

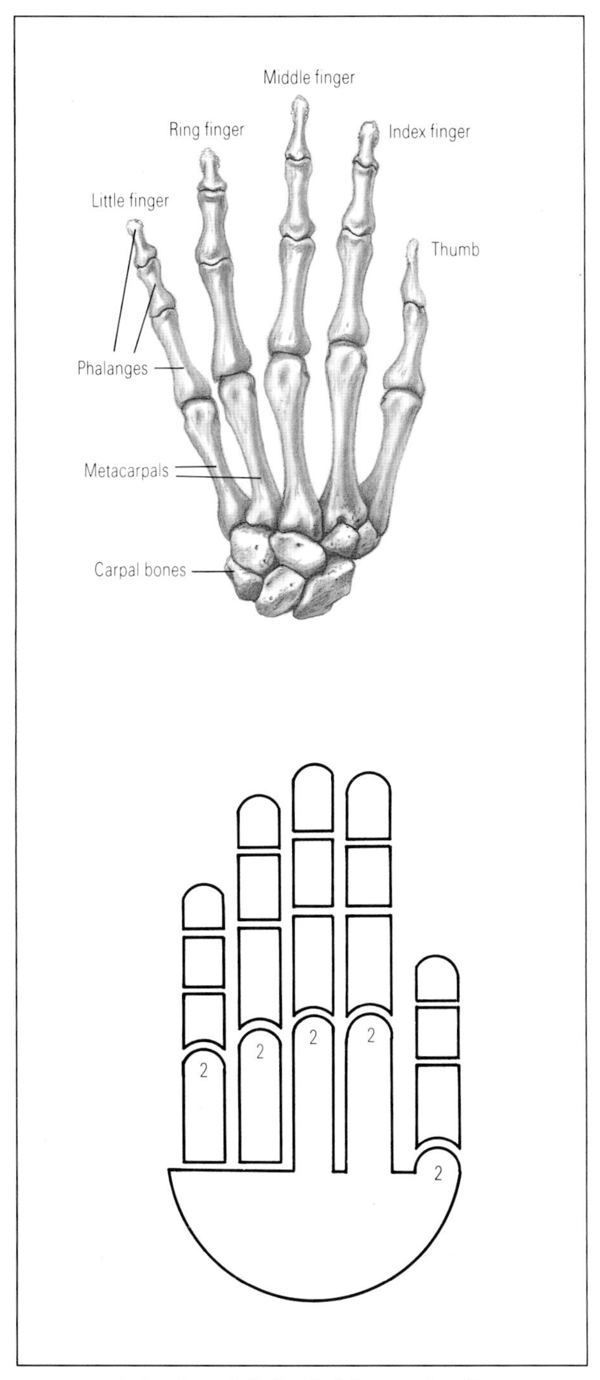

FIG. 2.5 *A drawing of the hand skeleton and a diagram showing joints with one or (labelled 2) two degrees of freedom.*

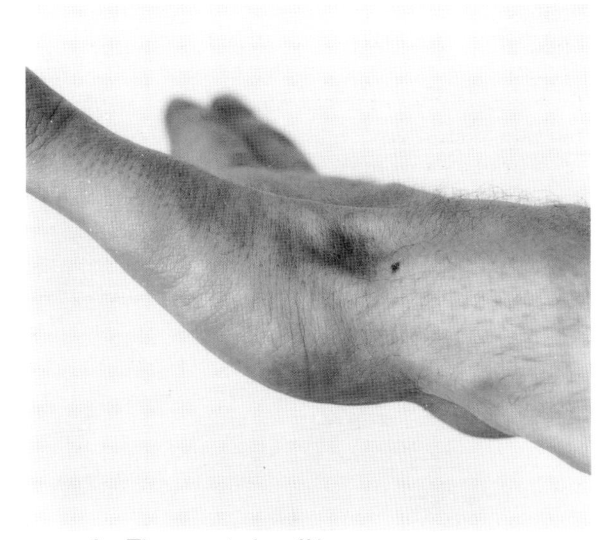

FIG. 2.6 *The anatomical snuff box.*

the 'anatomical snuff box' because snuff takers put a pinch of snuff into it, which they sniff up from there. Also, if you make a tight fist and feel the palm side of your forearm, just above the wrist, you can feel the tendons of the muscles that flex your fingers.

FIG. 2.7 shows these and other muscles of a finger. The superficial and deep flexor muscles (whose tendons we have just felt) and the extensor muscle are in the forearm, but the interosseous muscles lie between the metacarpal bones, in the palm of the hand. The tendons, running through the hands and along the fingers, pass under bands of collagen fibres that hold them close to the joints (FIG. 2.14b). Otherwise they would bowstring out from the bones when we bent our fingers.

The deep digital flexor tendon runs all the way from the forearm to the third phalanx (the bone of the finger tip) crossing on its way the wrist, the joint between metacarpal and first phalanx and the two joints between successive phalanges. Contraction of its muscles affects all these joints. The superficial digital flexor ends one bone earlier, on the second instead of the third phalanx. Its tendon forks just before attaching to the phalanx, allowing the deep flexor tendon to pass through. The extensor tendon runs over the back of the hand and along the finger, attaching both to the second and to the third phalanx. Apart from the tiny lumbricalis muscle (not illustrated) the two flexors and the extensor are the only muscles that cross the two hinge joints of the finger: here are three muscles, the fewest that can do the job, working joints that allow two degrees of freedom of movement. The interosseous muscles lie on either side of the metacarpal bone, and help to work the universal joint between it and the first phalanx. One swings the finger towards the thumb and the other towards the little finger, and the two acting together bend the metacarpo-phalangeal joint. Thus we have five muscles (two flexors, an extensor and two interossei) working a finger that has four degrees of freedom of movement (two hinges and a universal joint). Again, we have the minimum number of muscles that can do the job.

The superficial and deep flexor muscles each send separate tendons to each finger. Each tendon is pulled on by a separate group of muscle fibres so we can bend each finger separately, but it is difficult to bend the little finger without also bending the ring finger because there are connections between their deep flexor tendons.

The extensor muscle of the fingers has distinct parts with separate tendons for most of the fingers, but the extensor tendon of the ring finger also sends a branch to the little finger (FIG. 2.8). To discover how this limits the movements you can make, put your hands on the table as shown in FIG. 2.9. The right hand is flat on the table but the left one has the little finger bent, tucked under the palm. With your hands like this you can raise each finger of the right hand from the table in turn, but it is very difficult to raise the ring finger of your left hand (except by lifting it with the other hand). The branch of the ring finger extensor, that goes to the little finger, must be slack in the right hand, allowing the ring finger to be lifted without the little one. However, in the left hand, because the little finger is bent it prevents the muscle from shortening enough to lift the ring finger more than a very little.

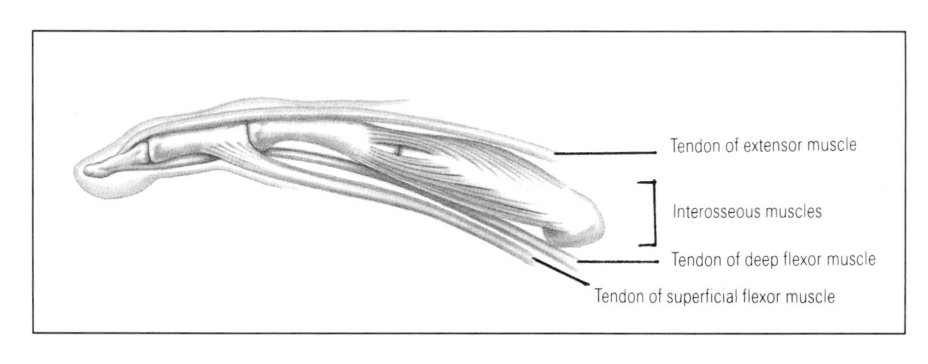

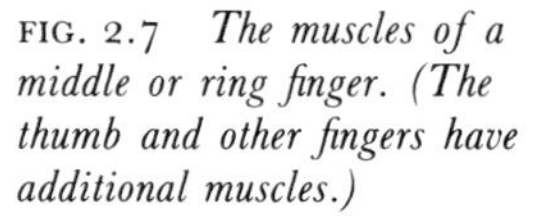
FIG. 2.7 *The muscles of a middle or ring finger. (The thumb and other fingers have additional muscles.)*

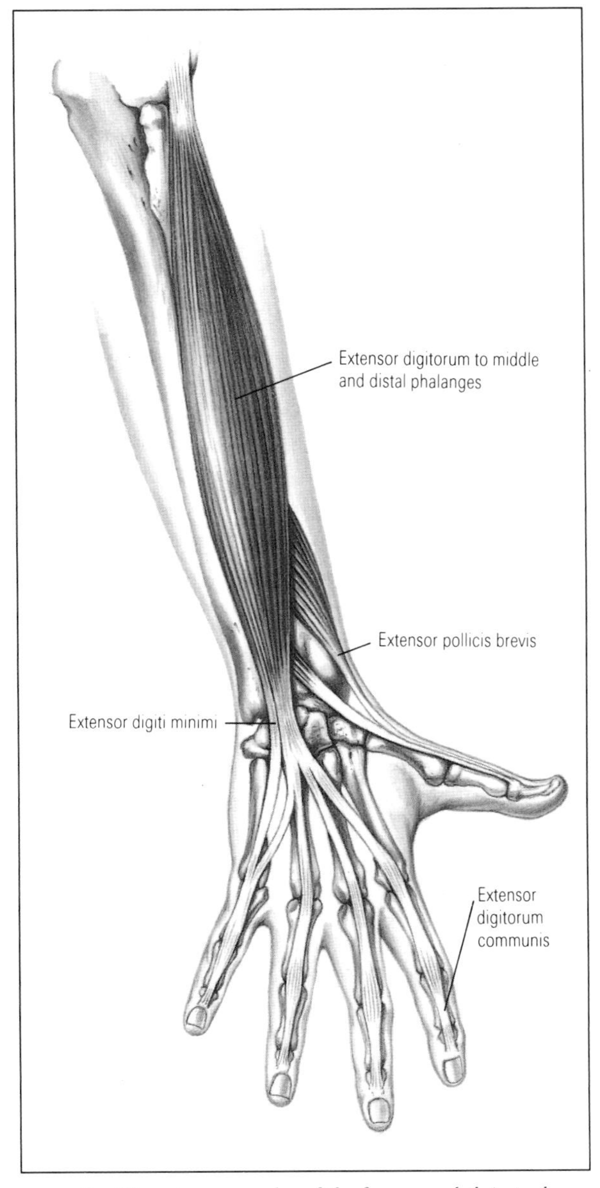

FIG. 2.8 *The extensor muscles of the fingers, and their tendons.*

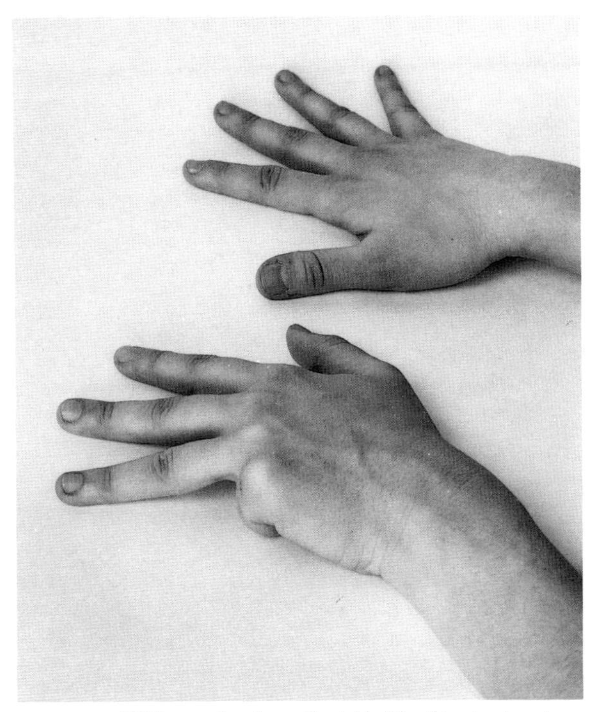

FIG. 2.9 *With your hands on the table like this, try to raise the ring finger of your left hand.*

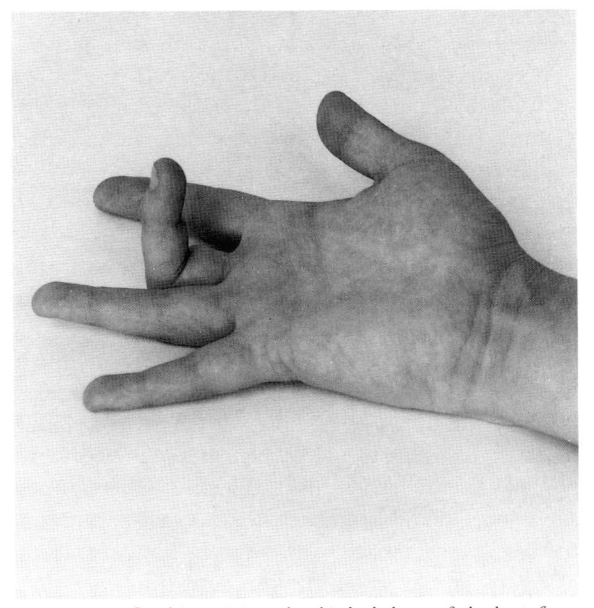

FIG. 2.10 *In this position, the third phalanx of the bent finger is out of control.*

You can discover another limitation in the control of your hand by bending only the middle joint of a finger, the joint between the first and second phalanges (FIG. 2.10). You will find that with the hand in this position you cannot hold the joint nearest the finger tip stiffly straight: if you manipulate it with the other hand you will find that it wobbles freely, totally out of control. The reason is that in this position the extensor muscle is prevented from shortening by its attachment to the second

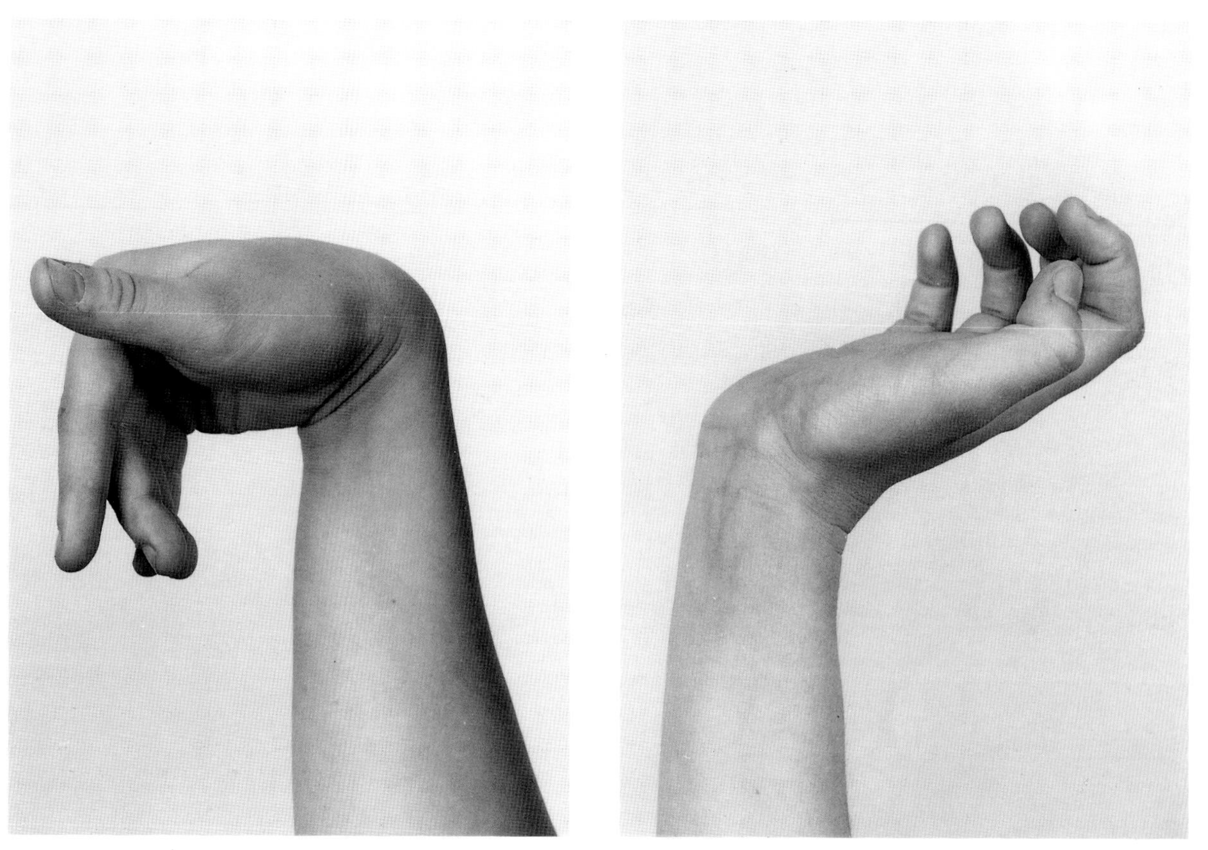

FIG. 2.11 *You cannot clench your fist while keeping the wrist strongly flexed (left) or straighten your fingers while the wrist is hyper-extended (right).*

phalanx, so the branches of its tendon that go to the third phalanx are slack and cannot be tightened.

It is said that you can make an assassin drop his knife by pushing on the back of his hand, forcibly bending his wrist. To show how this works, bend your wrist to a right angle as on the left in FIG. 2.11 and (keeping it bent) try to clench your fist tightly. I do not think that you will succeed. Now bend your wrist the other way (this is called hyper-extension) as on the right and try to straighten your fingers fully. Again, I do not think that you will succeed. To understand these limitations of hand movement we need to know something about how muscles work.

Under an ordinary microscope, fibres from hand muscles (or from any of the other muscles that work our skeletons) look striped. Electron microscopes show more detail, showing that the stripes reflect a regular arrangement of two kinds of filament: thick filaments of the protein myosin and thin ones that consist largely of the protein actin (FIG. 2.12).

These filaments do not change length as the muscle lengthens and shortens: rather, they slide between each other. Similarly, the segments of a telescope slide over each other without changing length. The muscle's force is exerted by cross bridges, projections from the thick filaments that attach to the thin ones and pull on them. Each cross bridge attaches, pulls, detaches and re-attaches further along the thin filament, working rather like someone pulling in a rope hand over hand.

When the muscle is extended as much as in FIG. 2.13(1), the thick and thin filaments barely overlap. Few of the cross bridges can attach and the muscle can exert little force. A wrapping of fine collagen fibres around the bundles of muscle fibres protects it from being stretched further. At the shorter length shown in FIG. 2.13(2), all the cross-bridges can attach and the muscle can exert maximum force. At a still shorter length (FIG. 2.13(3)) the ends of the thin filaments reach beyond the mid points of the thick ones to cross bridges that will push on them instead of pulling, reducing the force that the muscle

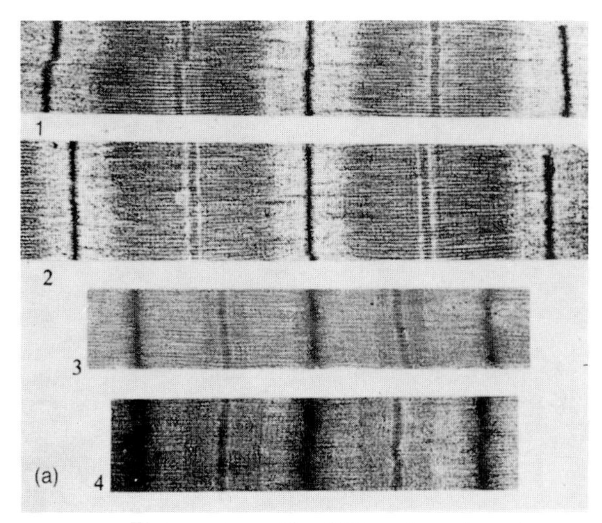

FIG. 2.12 *Electron micrographs of thin sections of muscle.*

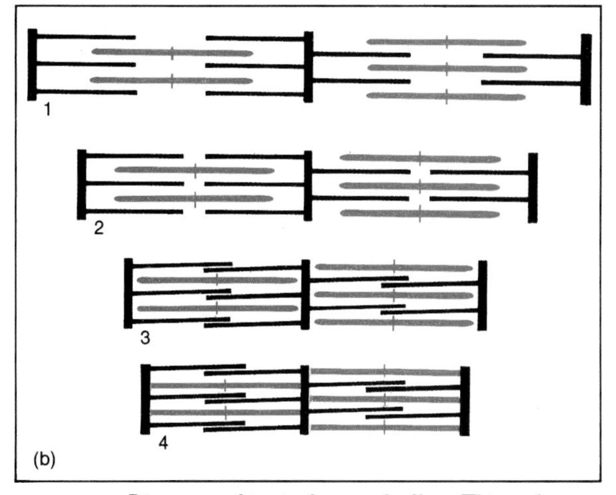

FIG. 2.13 *Diagrams of part of a muscle fibre. This and* FIG. 2.12 *from W.F. Harrington,* Muscle Contraction, *Carolina Biological Supply Company, 1981.*

can exert. Muscles cannot shorten much more than this because the thick filaments would soon be stopped by the Z discs, partitions that run across the fibre. Thus muscles can work only over a limited range of lengths and can exert maximum forces only in the mid part of the range. Flexing your wrist as in FIG. 2.11 (left) stretches the digital extensor muscles and slackens the digital flexors. To clench your fist with the wrist so bent you would have to stretch the extensors and shorten the flexors still more. That would apparently take one or both beyond their working range. Similarly, hyper-extending the wrist (FIG. 2.11, right) slackens the digital extensors and stretches the flexors making further shortening and stretching, respectively, difficult.

We will calculate the length changes involved for one particular muscle. When the muscle in FIG. 2.14a contracts, a point on the tendon alongside the joint moves along an arc of radius r (we call r the moment arm of the muscle). The circumference of a circle of this radius is $2\pi r$ (π is the number 3.14) or about $6.3r$. To rotate the joint through a complete revolution, which is of course impossible, the muscle would have to shorten by $6.3r$. To rotate it through a right angle it would have to shorten one quarter as much, by about $1.6r$, and proportionately for other angles. For finger muscles which cross several joints, length changes required for movements at the various joints add together. Consider for example the deep flexor muscle of the index finger (FIG. 2.14b). Its moment arm at the wrist in an adult man is about 17 millimetres, so to remain tight when the wrist bends through a right angle, it would have to shorten by $1.6 \times 17 = 27$ millimetres. Its moment arms at the

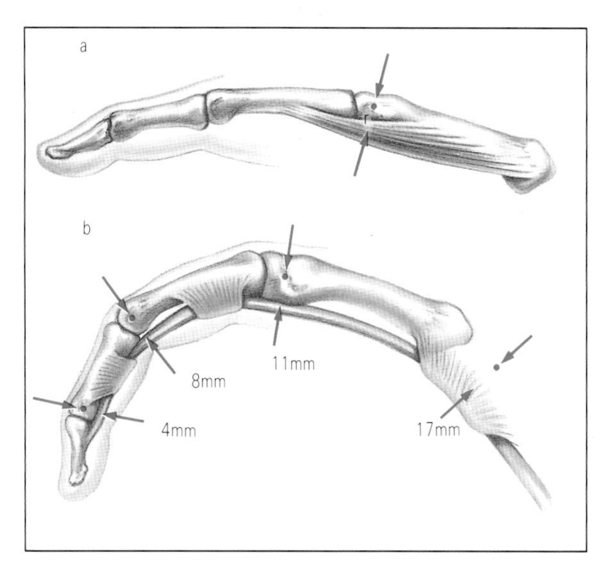

FIG. 2.14 *(a) A muscle that crosses just one joint. (b) The deep flexor muscle of a finger.*

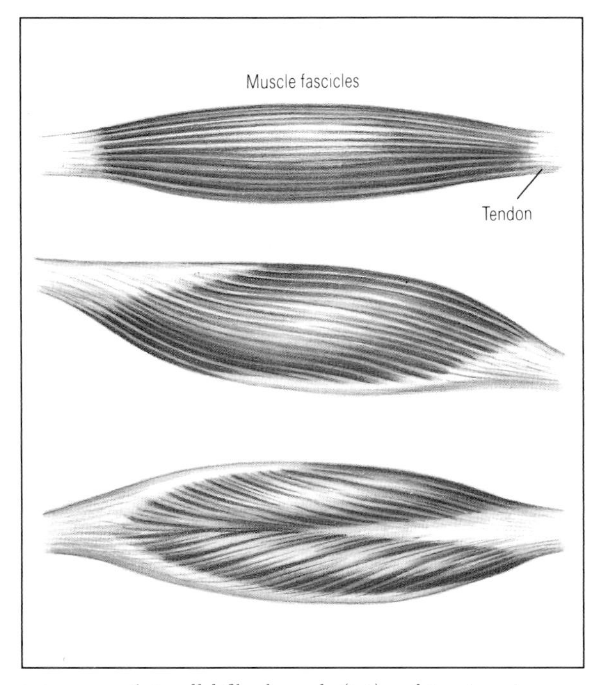

FIG. 2.15 *A parallel fibred muscle (top) and two pennate ones.*

metacarpo-phalangeal joint (at the base of the finger) and the two interphalangeal joints (in the finger) are about 11, 8 and 4 millimetres, respectively. To bend each of these joints through a right angle (which is roughly what happens when you make a fist) it would have to shorten 1.6 (11 + 8 + 4) = 37 millimetres. To make a fist while also bending the wrist to a right angle it would have to shorten 27 + 37 = 64 millimetres, from its length with wrist and fingers straight. This seems to be more than it can do.

The muscle consists of fascicles (bundles of muscle fibres) which are connected at one end to the membrane between the radius and ulna, and at the other to the long tendon that runs down the finger. Measurements of dissections show that these fascicles are about 100 mm long, when the wrist and fingers are straight. To clench the fist with the wrist still straight the fascicles would have to shorten by 37 mm, from their original 100 mm down to 63 mm. To clench it with the wrist flexed (which they cannot do) they would have to shorten by a further 27 mm to 36 mm, little more than one third of their original length.

Try squeezing a handy object such as a tennis ball in your fist, with your wrist bent to various angles. You will probably be able to squeeze hardest with your wrist fairly straight. The overlap between the thick and thin filaments in the fibres of the finger flexor muscles is presumably optimal (as in FIG. 2.13(2)) in that position.

We could of course have evolved finger muscles with longer fascicles that would have allowed us to clench our fists with our wrists bent, but if the number of fascicles were kept the same (to exert the same force) the muscle would be larger. The fascicle length that we actually have is presumably a compromise between bulk and mobility.

Some muscles (called parallel fibred) have their fascicles running the whole length of the fleshy belly and others (called pennate) have fascicles running obliquely to attach to their tendons (FIG. 2.15). Most of the muscles of our hands are pennate, with fascicles considerably shorter than their fleshy bellies.

FIG. 2.16 shows the hands of a squirrel. Hands like this seem to have appeared very early in the evolution of mammals and are found even on very primitive mammals such as the North American opossum. These primitive mammalian hands have five fingers like ours, the same number of bones, and joints allowing the same number of degrees of freedom. They have about the same number of hand muscles as we have (exactly the same in the case of the opossum) but are nevertheless much less dextrous. Squirrels can hold things only by curling their fingers around them with the 'thumb' behaving as just one more finger.

The most important difference between these primitive hands and our own is that we have a very mobile thumb. The heads of the metacarpal bones at the bases of the four fingers are bound together by a ligament that allows these metacarpals very little freedom to move, except as a group. However, the ligament does not extend to the first metacarpal (the one in the thumb) so it can move much more freely, relative to the others. Its range of movement is especially wide because its universal joint is at the

FIG. 2.16 *A squirrel's hands.* © *Planet Earth Pictures/Mark Mattock.*

base of the bone, not at the outer end as on the other metacarpals (FIG. 2.5).

The mobility of the thumb enables us to grasp things between it and the fingers, either in the precision grip that we use for pens and scalpels or in the power grip that we use for hammers and daggers (FIG. 2.17). The two degrees of freedom of the thumb, in the artificial hand of FIG. 2.4, make both precision and power grips possible. To appreciate the importance of thumbs, try performing a few simple tasks with your fingers alone. Pick up a notebook, open it at a chosen page and write your name, all without using your thumb.

Most of the ways in which we hold things can be classified either as precision grips or power grips, and involve the thumb, but there are two less important grips that use the fingers alone. These are the hook grip, used for carrying suitcases, and the scissors grip, which is generally used only for holding cigarettes.

The precision grip involves opposing the thumb to one or more of the fingers, bringing thumb and finger together pulp to pulp. That is something squirrels cannot do, but Old World monkeys can. For example, look at baboons picking parasites off each other (FIG. 2.18). The thumbs of apes have the same range of angular movement but are less effective for precision grips because they are much shorter than the fingers.

Precision grips enable us to move objects very delicately by finger movements alone. I can steady my hand by resting it on the table while I make very precise drawings. However, all the points of contact of hand with pen are close together so I cannot exert large moments to prevent the pen from being rotated if you grab it in a power grip and wrest it from my hand. In the power grip, there are well-separated points of contact and the object can be held firmly, but it is held against the palm and can only be moved by moving the hand as a whole. If you held

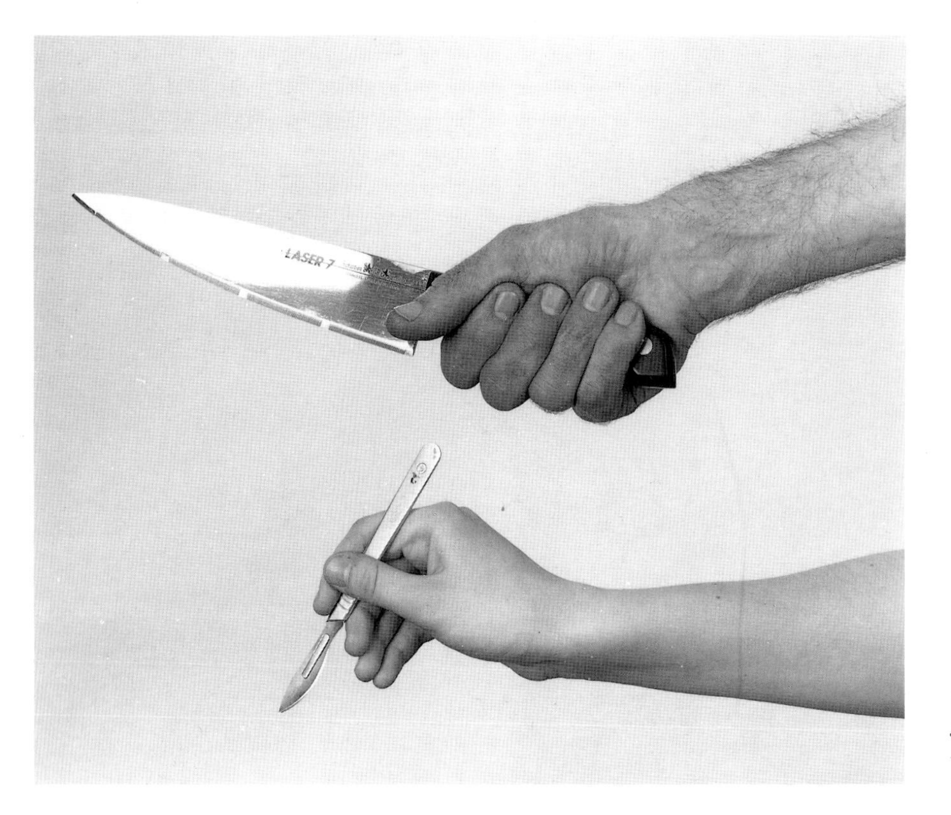

FIG. 2.17 *A large knife is held in a power grip and a scalpel in a precision grip.*

FIG. 2.18 *One baboon grooming another. © Dr Norman Myers, Bruce Coleman Ltd.*

your hammer in a precision grip you would strike a feeble blow and if you held your pen in a power grip you would write a clumsy scrawl.

Some grips make no use of friction, for example, when you hook your fingers under a suitcase handle. Others, such as the grip on the book in FIG. 2.2a depend on friction. We need to understand what friction is. You have to push things with an appreciable force to make them slide across a table. The force that resists your push is friction.

A simple experiment will show some of its properties. Take two identical glasses, one three-quarters full of water (not too full, because we are going to tilt it) and the other empty. Make sure both are dry on the outside and put them side by side on a tray. Push each with a finger and you will find (not surprisingly) that you have to push harder on the full glass to move it. The bottom of the full glass is pushed down harder on the tray by the weight of the water, which increases the friction. Now tilt the tray gradually until the glasses move. You will find that they start sliding at about the same angle of tilt. To see what that means, look at FIG. 2.19. The weight of the glass acts vertically downwards and must be balanced by the forces exerted on it by the tray. It is convenient to think of the tray exerting a 'normal' component of force at right angles to the tray surface (in this context 'normal' means perpendicular) and a frictional component of force parallel to the tray. We can discover the relative sizes of these forces by

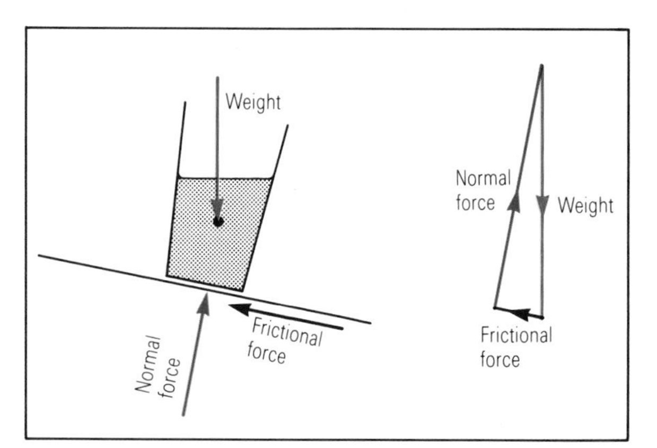

FIG. 2.19 *Forces on a glass on a tilted surface.*

drawing the triangle of forces, shown on the right of the figure. The principle of triangles of forces tells us that if a glass is in equilibrium under its weight and the normal and frictional forces, and if a triangle is drawn with each side parallel to one of the forces (the directions of the forces must either all point clockwise or all point anticlockwise round the triangle), the forces must be proportional to the lengths of the sides of the triangle. In this particular case the 'normal force' side of the triangle is five times as long as the 'friction force' side, so the friction force is 0.2 times the normal force.

FIG. 2.19 shows a tilt of 11 degrees, the angle at which I found glasses would just remain without sliding on one particular tray. I found that full and empty glasses both started sliding at about this angle, so the diagram will serve for either. It shows that the ratio of frictional force to normal force (measured just before sliding started) was about 0.2 in each case. This ratio is called the coefficient of friction. There is a general rule that it is almost constant for any particular pair of surfaces, whether they are being pressed together by small forces or by large ones. It is of course different for different surfaces: it is higher for shoes on dry concrete than on ice.

Barefoot people can squat on wooden boards tilted to a maximum of about 40 degrees, without sliding off. From that you can calculate that the coefficient of friction between feet and wood is about 0.8. I have used a more sophisticated (but still very simple) method to measure the coefficient of friction of the skin of my hands. I used a force plate, an instrumented panel that gives electrical outputs proportional to the forces that act on it: separate outputs give the vertical, forward-and-back and side-to-side components of the force. I pressed down on the plate with my finger and drew it across the plate so that the horizontal component registered by the equipment was the frictional force and the vertical component was the normal force. The ratio, the coefficient of friction, was different for different parts of the hand: it ranged from 0.3 for the skin of my knuckles to 1.0 for my finger tips (the palm side of the finger tips, where the fingerprint patterns are).

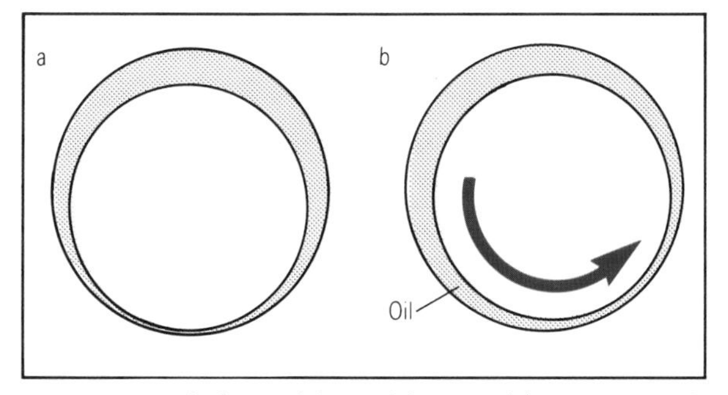

FIG. 2.20 *A shaft in a lubricated bearing, (a) stationary and (b) rotating.*

Those values are for sliding over the smooth (but not polished) aluminium top of the force plate, and I would presumably have got other values if I had covered the plate with some other material, or if my hands had been sweaty.

Holding a book even in the easier way (FIG. 2.2b) requires quite a hard squeeze. Even if the coefficient of friction is as high as 1.0, the normal (squeeze) force must equal the weight of the book.

Friction at our finger tips helps us to grip things but friction in our joints would make them stiff. An old bicycle will go much better after the moving parts have been oiled and, similarly, our joints need lubrication.

Lubrication depends on keeping a layer of oil or other lubricant between any surfaces that slide over each other – for example, between a rotating shaft and its bearings. When you put a shaft into an oily bearing, its weight squeezes out the oil from under it so that metal rests almost directly on metal (FIG. 2.20a). The stationary shaft is not well lubricated, but when it rotates, it drags oil along with it so that there is a thin layer of oil between the metal surfaces (FIG. 2.20b). This is called hydrodynamic lubrication. The coefficient of friction between dry metal surfaces is generally about 0.2, but hydrodynamic lubrication can reduce it below 0.01, a tremendous improvement.

When the shaft stops rotating it does not immediately sink onto its bearing, because it takes a little time for the viscous oil to be squeezed out from under it. Thus for a very short time it remains well lubricated and friction will still be low if it starts moving again, within this time. This is called squeeze film lubrication.

Hydrodynamic and squeeze film lubrication depend on keeping a film of lubricant between the moving surfaces. This film must not be too thin because no surfaces are perfectly smooth: if the film is too thin, the high spots on each surface will scrape against the other. Hydrodynamic lubrication depends on the viscosity of the lubricant to pull it round with the rotating shaft. (Viscosity is the property that makes treacle hard to stir.) Squeeze film lubrication depends on the viscosity of the lubricant to delay its squeezing out. In both cases, the lubricant must be sufficiently viscous to maintain the necessary film thickness.

Once metal surfaces have been oiled, friction between them remains less than between clean metal surfaces, even when all the oil seems to have drained away. The reason is that a single layer of oil molecules remains firmly attached to the metal surfaces, preventing them from rubbing directly against each other. This is called boundary lubrication. It gives coefficients of friction of (typically) about 0.05.

Human joints do not have rigid surfaces, like axles and bearings in machinery. Instead they are covered by a layer of cartilage, one or two millimetres thick on the joints of the fingers but five millimetres or even more on larger joints such as hips and knees of healthy young people. This is a translucent substance with rubber-like elastic properties. It has living cells in it, but they occupy only a small fraction of its volume. The rest consists typically of around 20% collagen fibres, 5% proteoglycans and 75% water. Collagen is the protein that we have already found in tendons and ligaments. Its fibres have their ends firmly attached to the bone but loop up close to the surface of the cartilage (FIG. 2.21, inset). Proteoglycans are large molecules that combine protein with polysaccharides (sugar polymers). Electron microscope pictures show polysaccharide chains attached to a protein core, like the needles on the twigs of a Christmas tree. These proteoglycans are attached to long flexible molecules of hyaluronic acid, another polysaccharide. The proteoglycans have the property of drawing in water, making the cartilage swell, while the collagen fibres prevent it from swelling too much.

The cavity of the joint is filled with synovial fluid, which is mainly water but also contains hyaluronic acid (which makes it viscous) and protein. Its properties are complicated but, to give a very rough indication, its viscosity lies in the range spanned by lubricating oils. It must lubricate the joint, and it is prevented from draining away by a membrane (the joint capsule) that encloses the whole joint.

Many experiments have shown that the lubrication is very effective. Most of them have been done on joints from dead animals but one set of experiments was done on the fingers of living people. I described earlier in this chapter how the joint nearest the tip of a finger can be released from the control of its muscles, by bending the finger's middle joint. A person's hand was strapped palm up to a table top, in the position of FIG. 2.10, and a pendulum was suspended from the tip of the middle finger so that

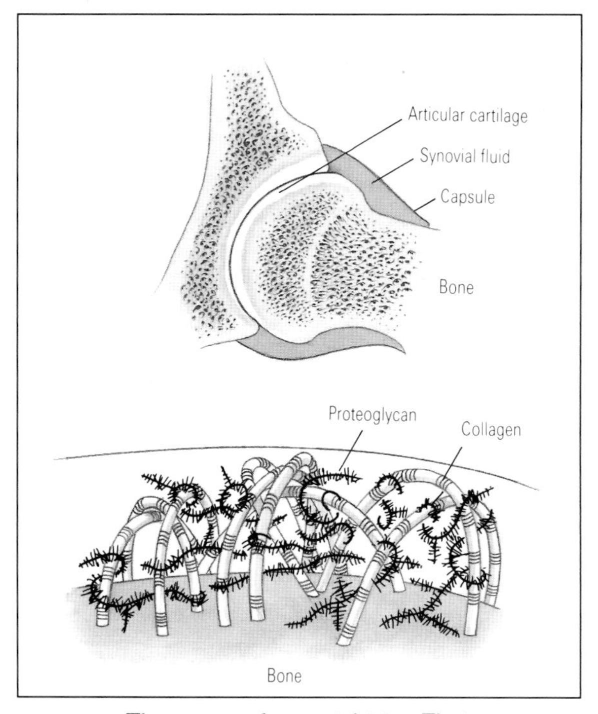

FIG. 2.21 *The structure of a synovial joint. The inset represents part of the articular cartilage, greatly enlarged.*

the released joint served as the pendulum's pivot. The pendulum was set swinging and the coefficient of friction of the joint was calculated from the rate at which the oscillations died away. Values around 0.008 were obtained, as low as for good engineering joints.

Hydrodynamic lubrication is very effective in machinery, in which shafts generally rotate continuously in one direction. It cannot work so well in human joints which never keep turning in one direction for more than a small fraction of a revolution, and are constantly stopping and starting. Boundary lubrication could work in such circumstances, but human and animal joints have much lower coefficients of friction than boundary lubrication can be expected to give. Squeeze film lubrication is probably important in activities such as walking in which the load is taken off the joints once in each step, allowing the cartilage surfaces to separate and synovial fluid to flow in between them.

We do not fully understand why human and animal joints work so well. One important insight came from Charles McCutchen, a scientist with an extraordinary talent for unconventional experiments. He realised that cartilage is rather like a sponge: it soaks up fluid, which can be squeezed out again. He and a colleague suggested that this sponginess might make possible a form of lubrication – weeping lubrication – that is unknown in engineering.

To demonstrate the principle they took a piece of sponge rubber (the sort that contains individual bubbles of air rather than a network of cavities) and cut it to expose a layer of bubbles. They measured the coefficient of friction between the cut surface and glass, using soapy water as a lubricant, and obtained very low values, around 0.003. The reason it is so low seems to be that when the rubber is pressed against the glass, nearly all the load is borne by the water in the cut-open bubbles. The rubber that encloses the bubbles makes contact with the glass, preventing the soapy water from escaping sideways, but because it deforms so easily it bears very little of the load. However, the trapped water does escape slowly and the coefficient of friction rises if the rubber is kept pressed hard against the glass.

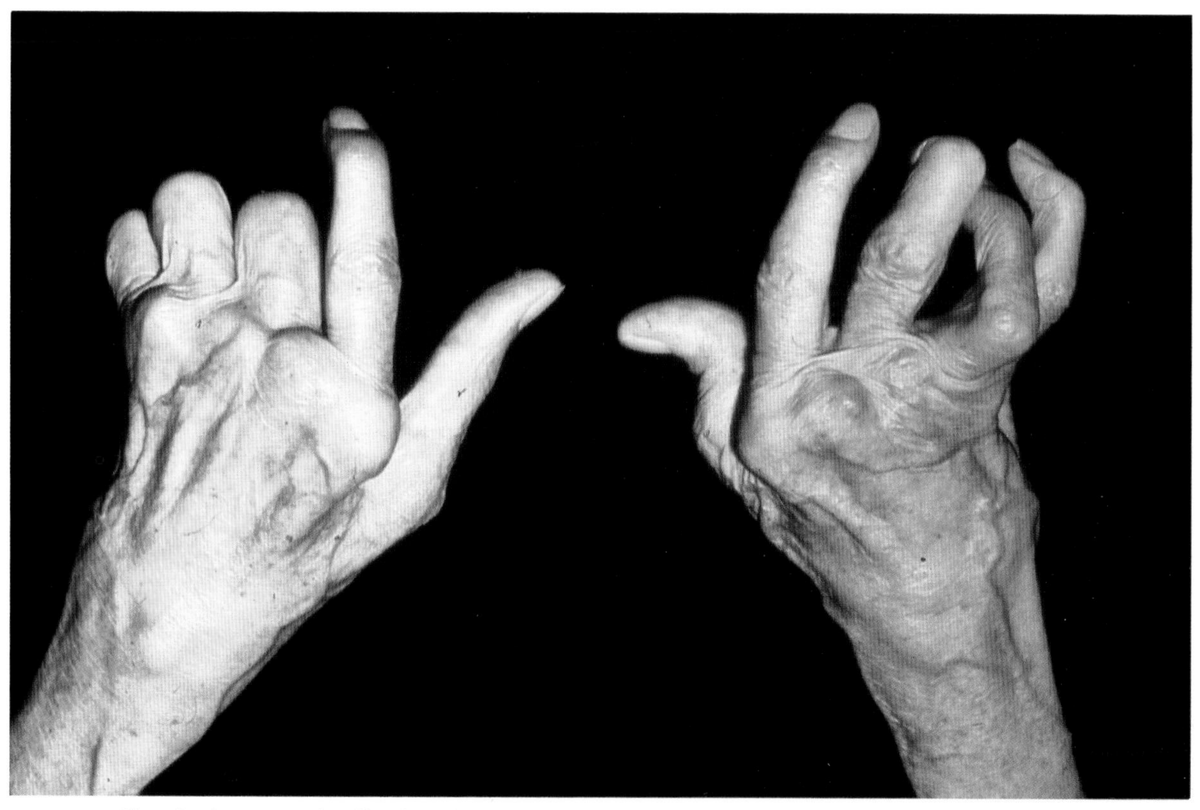

FIG. 2.22 *These hands are severely affected by rheumatoid arthritis. Photograph supplied by Professor V. Wright, Department of Rheumatology, University of Leeds.*

Cartilage is really more like a bath sponge than the rubber used in the experiment, so fluid will be squeezed out sideways through the cartilage itself, as well as between the cartilage surfaces. However, the pores are very fine so the squeezing out will happen slowly. The cartilage is squeezed gradually thinner, if the joint is kept heavily loaded for several minutes, and the coefficient of friction rises to values typical of boundary lubrication. If the joint is relieved of the load for a while the cartilage soaks up fluid again and the coefficient of friction falls to the original value. It has been suggested that one reason why we fidget while standing is to move the loads to different parts of our joints, allowing squeezed-out parts to recover.

Several important refinements to the theory of weeping lubrication have since been suggested. Cartilage becomes much less permeable to fluid as it is squeezed out, so it becomes increasingly difficult to squeeze further. The water from the synovial fluid may be squeezed out through the cartilage but the protein and carbohydrate molecules that make the fluid viscous cannot escape that way, so squeezing out leaves an increasingly concentrated layer of lubricant molecules between the cartilage surfaces. Also, the cartilage deforms under load, increasing the area in which the cartilage surfaces are brought close together. This slows down the flow of fluid between the cartilage surfaces and prolongs the squeeze film effect.

Unfortunately, the lubrication of our joints often breaks down, causing the various forms of arthritis. Osteoarthritis is damage by wear and tear. Rheumatoid arthritis is a disease whose cause remains a puzzle. In both, the articular cartilage of affected joints may be thinned or even lost, leaving bare bone surfaces that may become polished by wear. The joint becomes painful and stiff, and may emit creaking or grating noises when moved. The joints of the fingers are often affected and further degenerative changes may occur, resulting in a seriously deformed hand. (FIG. 2.22).

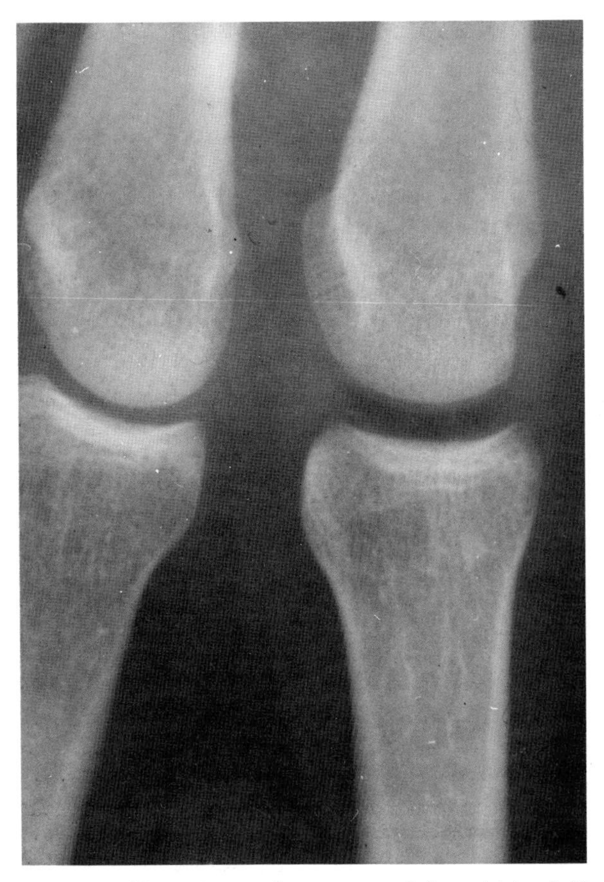

FIG. 2.23 *X-ray pictures of a metacarpo-phalangeal joint (left) before and (right) after cracking. These pictures were supplied by Professor A. Unsworth of the School of Engineering and Applied Science, University of Durham.*

X-ray pictures of normal joints show a gap, the articular cartilage, between the X-ray-opaque bones. This gap is often obviously narrowed in arthritic joints. X-rays of osteoarthritis often show bony outgrowths around the joint whereas those of rheumatoid arthritis may show the bone partly eaten away. The earliest sign of developing rheumatoid arthritis, however, is inflammation of the membrane that lines the joint capsule, which becomes swollen with excess synovial fluid. The symptoms of all kinds of arthritis can be relieved by appropriate drugs but it is often impossible to get the natural joint working satisfactorily again, and an artificial replacement has to be fitted (FIG. 4.2).

Osteoarthritis often develops for no obvious reason, especially in women over the age of fifty. In many other cases it seems to be due to over use of a joint in the course of work or sport, for example in the hands of textile workers and the knees of footballers. A group of clinicians examined the hands of 64 women, all of whom had been working in a U.S. textile mill for the past 20 years. These women were still working satisfactorily but many showed signs of osteoarthritis in some of their joints. Some of the women were winders, doing a job that involved a power grip with marked wrist movement but little finger movement. Their wrists suffered more from osteoarthritis than did those of the other workers. These were burlers and spinners, doing jobs that used precision grips, and suffered more from osteoarthritis in the joints of the fingers.

Some people can crack their fingers: a pull on the finger produces a sharp cracking noise from the joint at its base (the joint between the metacarpal and the first phalanx). X-ray pictures show a bubble of gas in the joint, after it has been cracked. This shows up as a dark area on the right of the picture (FIG. 2.23): just as bones show up as light patches in X-ray pictures because bone is more opaque to X-rays than other tissues, gas bubbles show up dark because they are more transparent. The joint cannot be cracked again until (after about twenty minutes) the bubble disappears.

This seems to be an example of the phenomenon that engineers call cavitation: the appearance of bubbles of vapour in water at low pressure. Synovial fluid cannot flow rapidly into the narrow space between the two bones that form the joint, so a brief pull on the finger produces a negative pressure in this small space. If the pressure falls low enough, some of the water in the space becomes vapour, greatly increasing its volume and allowing the bones to separate by about three millimetres. Fluid can now flow in rapidly and the bubbles of low-pressure vapour collapse, producing the cracking sound. The bubbles do not disappear completely at this stage because synovial fluid has air dissolved in it, and some of this air enters the vapour bubbles when they form. When the bubbles collapse this air remains, and the joint cannot be cracked again until it re-dissolves.

Even if you can crack your joints it may be wiser not to, in case it damages the cartilage.

CHAPTER THREE

LIFTING

So far we have considered only the movements that muscles can produce but this chapter is about the forces they exert, some of which are remarkably large. Consider, for example, me holding my briefcase at arm's length (FIG. 3.1). The briefcase had a mass of 5 kg (there were several books in it) and was quite difficult to hold in this position. We will calculate the force needed in the muscles of the shoulder joint. Two muscles are involved, the deltoid and the supraspinatus. To find out how much force they must exert we have to consider the weights both of the briefcase and of the arm.

At this point we need to be clear about the words 'mass' and 'weight'. If the forces acting on a motionless body are balanced, it will remain motionless. Alternatively, if balanced forces act on a body that is already moving, it will continue moving with the same velocity (that is, at the same speed in the same direction). Any unbalanced force would change its velocity. Acceleration is rate of change of velocity: for example, if a falling stone speeds up from 10 metres per second to 20 metres per second in a second, its acceleration is 10 metres per second squared (m/s^2). Newton's second law of motion tells us that

$$\text{force} = \text{mass} \times \text{acceleration}.$$

The scientific unit of force is called the newton. One newton is the force needed to give a mass of one kilogramme an acceleration of 1 m/s^2.

Scientists make a sharp distinction between mass and weight. The mass of an object (generally measured in kilogrammes) tells us about its inertia, about how hard it is to accelerate. Its weight (measured in newtons) is the force that gravity exerts on it. If you took an object to the moon its mass would remain the same but its weight would be less because gravity is weaker there. On earth, an object falling freely in a vacuum has an acceleration of 9.8 m/s^2. Therefore, by Newton's second law, a mass of 1 kilogramme has a weight of 9.8 newtons.

That brings us back to the briefcase. Its mass was 5 kilogrammes so its weight was $5 \times 9.8 = 49$ newtons. The mass of a typical man's arm is about 3.6 kilogrammes, corresponding to a weight of 35 newtons. These forces are shown in FIG. 3.1. We want to calculate their effects at the shoulder.

A small child sitting at one end of a seesaw can balance a heavier one sitting nearer the pivot on the other side. This is because the turning effect of a force is given by its moment, the force multiplied by the perpendicular distance from its line of action to the pivot. Multiplying the force in newtons by the distance in metres gives the moment in newton metres. In the example of FIG. 3.1, the 49 newton weight of the briefcase is acting 0.65 metres from the shoulder joint so its moment is $49 \times 0.65 = 32$ newton metres. The 35 newton weight of the arm acts at its centre of gravity which is 0.32 metres from the joint giving a moment of 11 newton metres. Thus the total anticlockwise moment at the joint is $32 + 11 = 43$ newton metres. To balance it, the shoulder muscles must exert an equal clockwise moment. Their moment arm is 25 millimetres (that means, their line of action is 25 mm = 0.025 metres from the joint) so the force they must exert is 43/0.025 or 1700 newtons (the weight of a $1700/9.8 = 175$ kg mass). Less force would be needed if the briefcase were held nearer the body, reducing its moment about the shoulder.

(FIG. 3.1 also shows a force that I have not men-

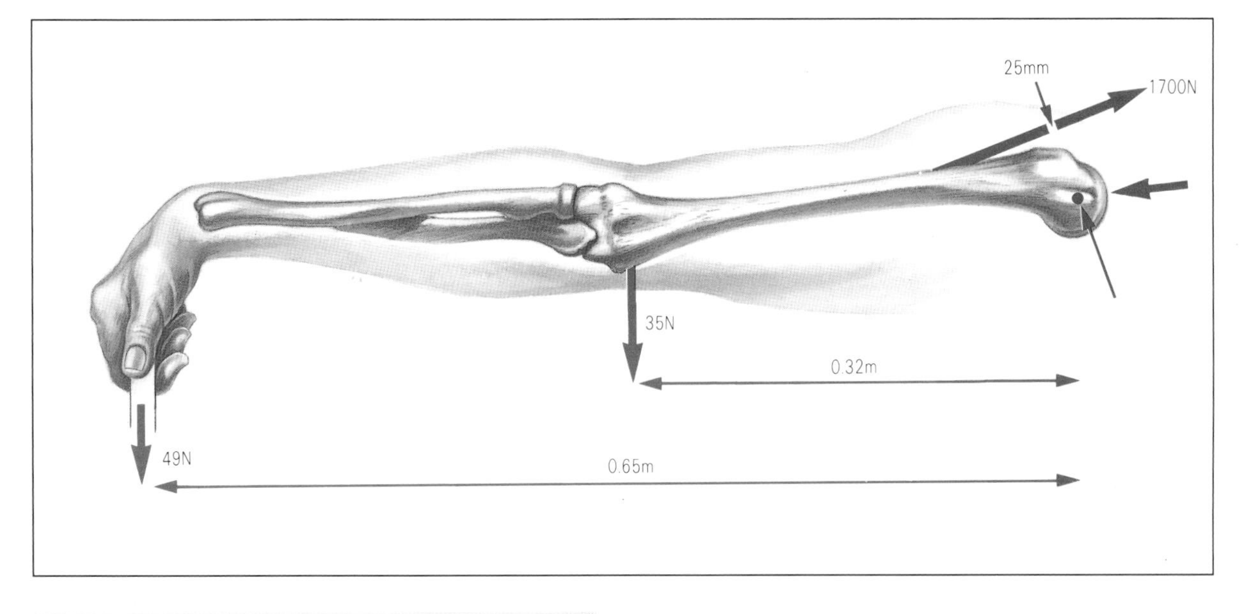

FIG. 3.1 *The author holding a briefcase (above) and (top) the forces that act on his arm.*

tioned, the reaction that the scapula exerts where it pushes against the head of the humerus. This force acts at the joint, so it has no moment about the joint, and we did not have to include it in the calculation.)

Even quite small muscles can exert substantial forces, for example the flexor muscle of the thumb. Unlike the fingers, the thumb has only one interphalangeal joint, and only one flexor muscle to bend it. This muscle was the subject of one of the earliest biomechanical calculations, in a book by the seventeenth-century Italian scientist Giovanni Borelli. I reproduce his diagram (FIG. 3.2) but will take the numbers for the calculation from much more recent research. Borelli found that a mass of 20 libbra (probably about 6.8 kilogrammes) could be supported as shown, from the tip of the thumb. The moment arm of the load about the joint was about three times that of the muscle so he estimated the force exerted by the muscle to be 60 libbra (about 20 kilogrammes force, or 200 newtons). In the more recent investigation, a force transducer (an electrical device for measuring forces) was connected to a person's thumb by wire and adhesive tape. The forces it registered, when the subject bent his thumb strongly, showed that the muscle was exerting a force of 130 newtons, one third less than Borelli's estimate.

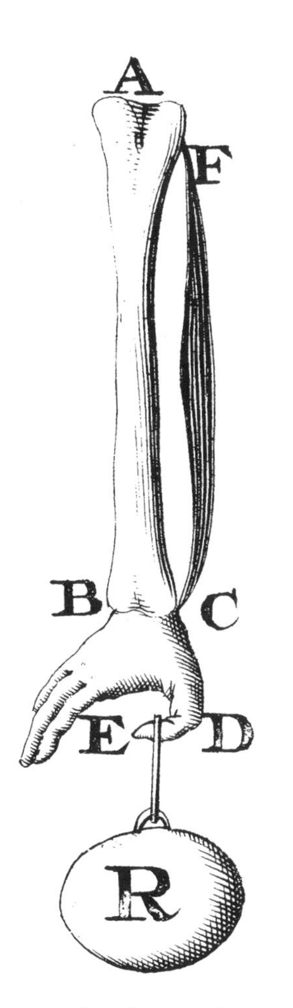

FIG. 3.2 *Giovanni Borelli's diagram of the flexor muscle of the thumb. From G.A. Borelli (1680)* De Motu Animalium *Bernabro, Rome.*

Thick ropes are stronger than thin ones. Similarly, we may expect the force that muscle can exert to depend on its cross-sectional area or, more precisely, on the total of the cross-sectional areas of all its fascicles. The flexor of the thumb is pennate (see FIG. 2.15) so no single section across it will cut every fascicle. We can discover the area we want in a different way. Alison Cutts and I dissected the thumb flexor muscle from an accidentally severed but otherwise healthy arm. The mass of the muscle was 27 grammes. Muscle is slightly heavier than water. Its density (mass divided by volume) is 1.06 grammes per cubic centimetre so the volume of the thumb flexor muscle must have been $27/1.06 = 25$ cubic centimetres. Its fascicles averaged 6 centimetres long (with wrist and thumb straight) so the total of their cross-sectional areas (volume divided by length) was 4.2 square centimetres. When the muscle was exerting its maximum force of 130 newtons the stress in its fascicles (force per unit cross-sectional area) was 31 newtons per square centimetre (0.31 newtons per square millimetre).

The forces that muscles can exert have been measured more directly, in experiments with muscles from small animals such as rats. A small muscle dissected from a freshly-dead animal can be kept alive for hours in a suitable solution and made to contract as required by electrical stimulation. Experiments like this with various muscles from rats, mice, frogs and other vertebrate animals have shown that the maximum stresses they can exert are about 0.3 newtons per square millimetre, about the same as for the human thumb flexor. That is the stress when the thick and thin filaments are overlapping well (as in FIG. 2.13(2)) and the muscle is holding constant length, neither lengthening nor shortening.

Muscles cannot exert such large stresses when they are shortening, and the faster they shorten the less stress they can exert: you cannot lift heavy weights quickly. FIG. 3.3 shows an experiment designed to measure this effect. A small muscle dissected from a freshly-dead animal is attached to an electrical device which can be used to stretch it or allow it to shorten at controlled rates. The muscle is kept active by electrical stimulation while its length is being changed, and a transducer records the force that it exerts. The results show that the force that the muscle can exert falls off rapidly as shortening speed increases, and that no force can be exerted above a certain speed: above this maximum speed the muscle cannot shorten fast enough to keep up with the apparatus. They also show that when the muscle is forcibly stretched it can exert very large forces, but only up to a limit. As a general rule, muscles that exert 0.3 newtons per square millimetre while holding constant length can exert up to 0.5 newtons per square millimetre while being forcibly stretched.

If there were nothing to prevent it, any stress of more than 0.3 newtons per square millimetre would

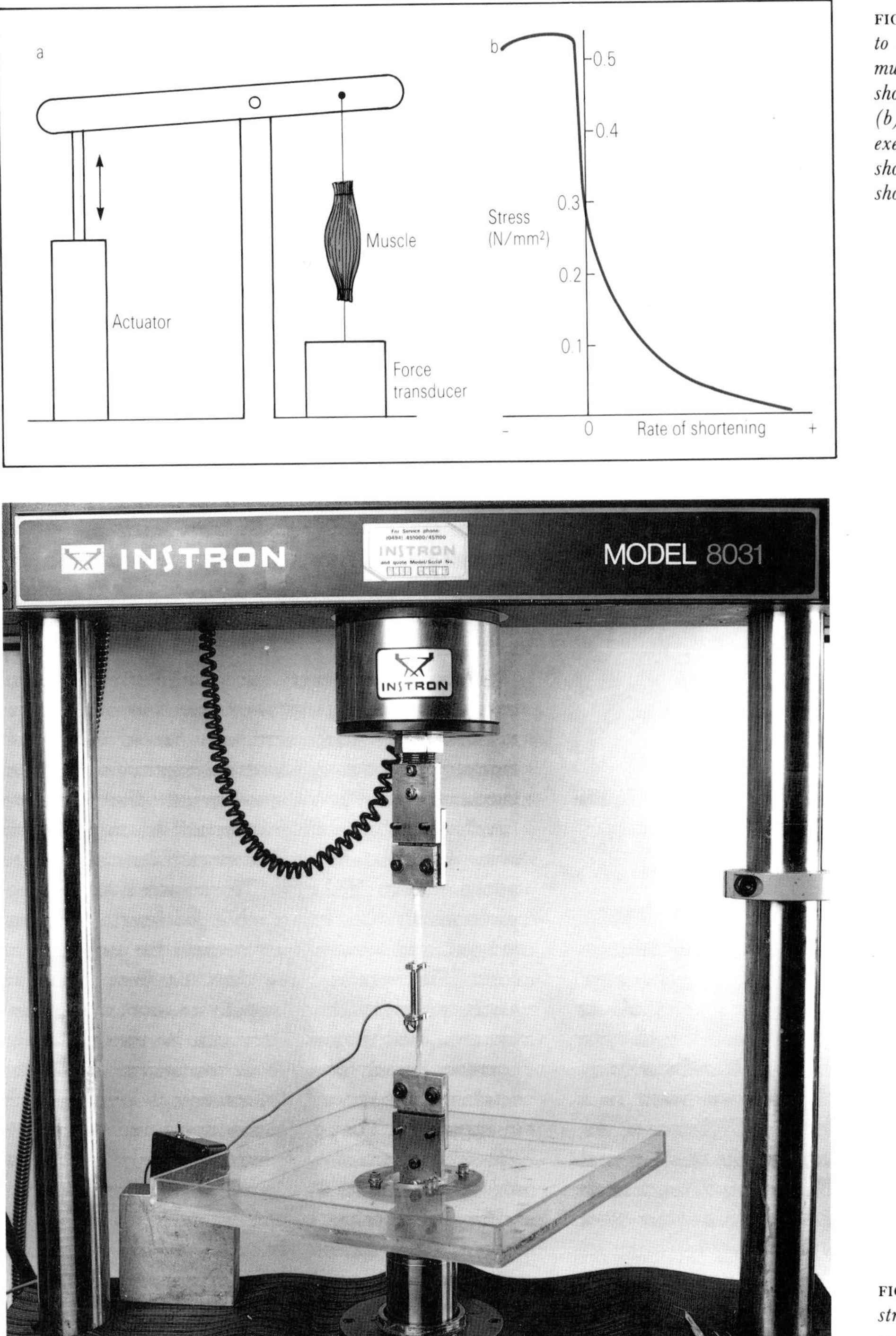

FIG. 3.3 *(a) An experiment to measure the forces a muscle can exert while shortening or lengthening. (b) A graph of the stress exerted against the rate of shortening. Negative shortening means stretching.*

FIG. 3.4 *A tendon being stretched in a testing machine.*

eventually make the thick filaments slide out from between the thin ones: the muscle would disintegrate. In practice this is prevented by sheaths of fine collagen fibres around the muscle, around each of its fascicles and around the individual fibres. These sheaths prevent the muscle from being stretched too far. However, excessive forces may tear them. Occasionally muscles tear right across but more often only a few of the fascicles are torn. Torn muscles are common injuries in athletes.

Our muscles can exert 0.3 newtons per square millimetre (when holding constant length) but the tensile strength of tendon (the stress needed to break it) is about 100 newtons per square millimetre. For that reason, a slender tendon can transmit the force of a stout muscle. The tendon of the human thumb flexor muscle has a cross-sectional area of 11 square millimetres so the stress in it, when the muscle exerts 130 newtons, is only 12 newtons per square millimetre. The tendon is much thicker than it needs to be, to take the muscle's force without breaking.

The reason it grows so thick is probably that a thinner tendon would stretch too much. Imagine that the cables that work your bicycle brakes were made of rubber instead of steel wire, so that they were stretched quite a lot as you applied force to the brakes. You would have to move the brake levers a long way, to apply much force, so you could not apply the brakes quickly. Indeed, the lever might come to the end of its travel before much force was developed, preventing you from braking hard. Imagine on the other hand that the brake cables were utterly inextensible. A tiny movement of your hand would make a huge difference to the braking force, and it would be very difficult to control your braking. A little compliance (that means elastic extensibility) in the brake cables is helpful but too much would be troublesome or even disastrous. Compliance in the tendons of hand muscles brings similar advantages and disadvantages.

The compliances of many tendons have been measured in machines of the kind that engineers use for testing metals and plastics (FIG. 3.4). The principle of the machine is shown in FIG. 3.5. The load cell is fixed at the top and gives an electrical output proportional to any force that acts on it. The hydraulically-driven actuator can be made to move up and down, exerting large forces if necessary. The

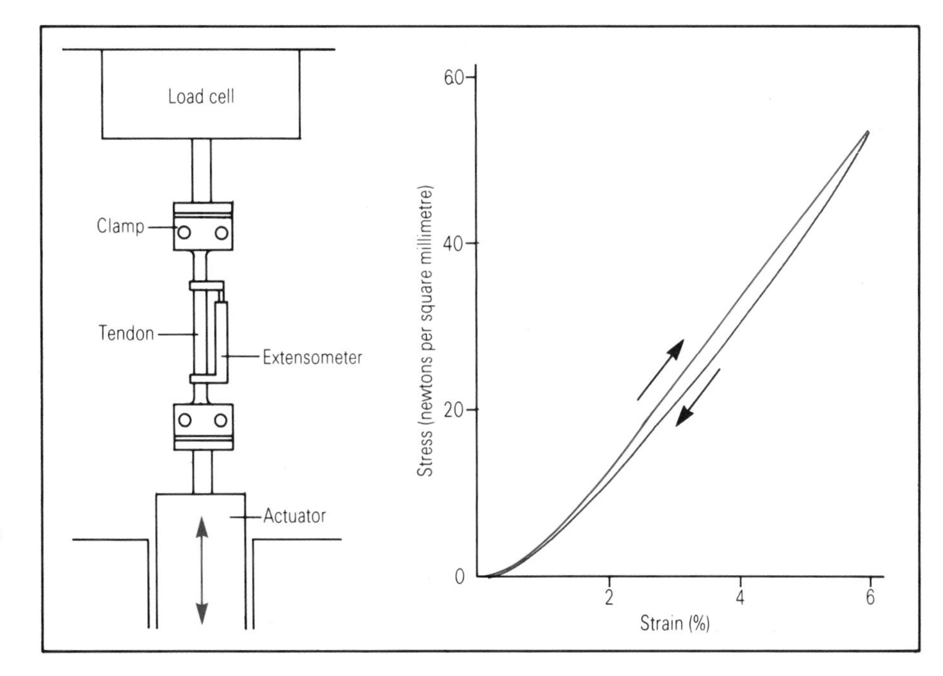

FIG. 3.5 *A diagram of the machine shown in* FIG. 3.4 *and the record of an experiment in which a human big toe flexor tendon was stretched and allowed to recoil. Each cycle of stretch and recoil lasted 2 seconds.*

tendon (which must be kept moist) is clamped at one end to the load cell and at the other to the actuator. It is stretched as the actuator moves down and recoils as it rises. The clamped ends of the tendon get squashed badly out of shape so it is best to use an extensometer to measure the length changes of an undistorted section of the tendon.

The graph in FIG. 3.5 is the record of a test on a human big toe flexor tendon. (Tests on other human and animal tendons give very similar results.) It shows that a stress of 12 newtons per square millimetre (the stress that we calculated for the thumb flexor tendon) stretched the tendon by 2%. The tendon of the thumb flexor is 120 millimetres long from its attachment to the bone to the beginning of the muscle's fleshy belly, so 2% extension means 2.4 millimetres lengthening for the free part of the tendon. The moment arm of the thumb flexor at the interphalangeal joint is 8 millimetres, and the calculated extension of the tendon is enough to allow the joint to move through 17 degrees. It is easy to imagine that a thinner tendon, which stretched more and allowed more movement, might be a disadvantage.

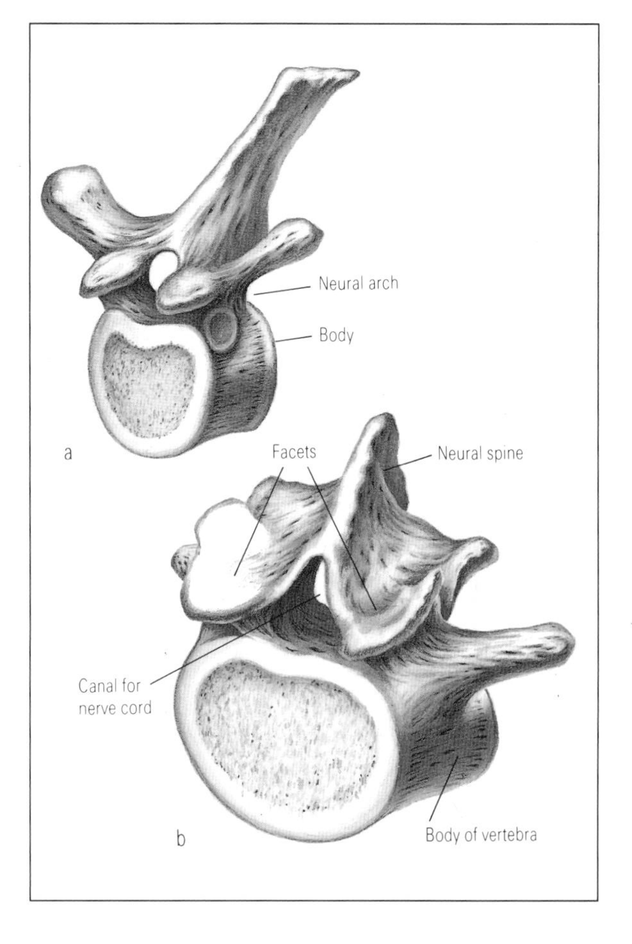

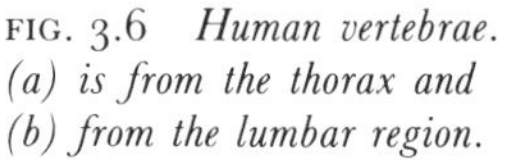
FIG. 3.6 *Human vertebrae. (a) is from the thorax and (b) from the lumbar region.*

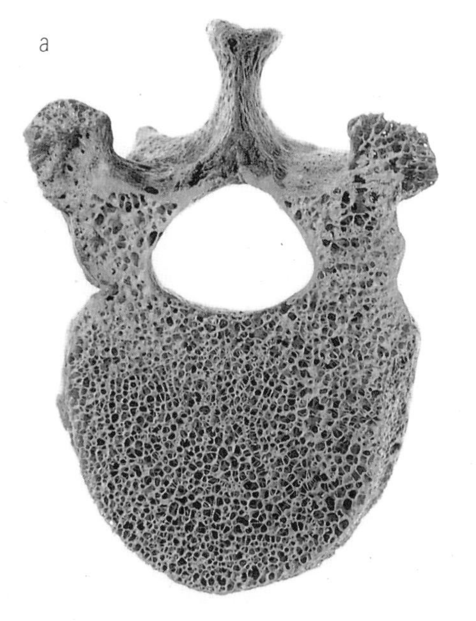

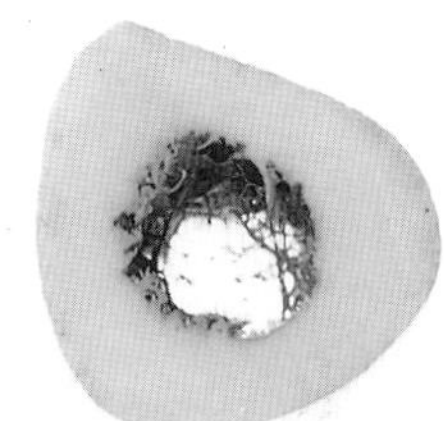

FIG. 3.7 *Sections through (a) the body of a vertebra and (b) the shaft of a long bone.*

To lift heavy loads we need strong arms, strong legs and strong backs. Quite a lot has already been said about arms, and legs feature prominently in later chapters. We will concentrate here on the mechanics of the back.

The spine is a chain of vertebrae. We have seven cervical vertebrae in our necks, twelve thoracic vertebrae (which bear ribs) in the upper back and five lumbar vertebrae (without ribs) in the lower back. Next is the sacrum, a group of five vertebrae fused into a single block of bone, which is firmly attached to the pelvic girdle, and finally the coccyx, all that remains of the tail of our monkey ancestors.

Each vertebra has a body and a neural arch (FIG. 3.6). The body is the main load-bearing part and the arch covers and protects the spinal cord. The bone of a complete vertebra looks solid, but cuts through it show that it is porous, permeated by spaces that are filled in life by marrow (FIG. 3.7a). In contrast, our principal arm and leg bones have walls of compact (not spongy) bone enclosing a central marrow-filled cavity (FIG. 3.7b). Each vertebral body is connected to the next by an intervertebral disc a few millimetres thick. The outer part of the disc consists mainly of collagen fibres arranged in criss-cross fashion but the centre (the pulp) is a semi-fluid jelly of proteoglycan in water. Loads on the vertebral column squeeze some of the water out of the pulp but the proteoglycans draw it in again when the load is removed. For this reason, adult people are about 20 millimetres taller when they get up in the morning, than when they go to bed at night.

The flexibility of the discs allows movement between neighbouring vertebrae, but the range of movement may be restricted by facets on the neural arch that rub against corresponding facets on the adjacent vertebrae. These facets are covered by articular cartilage, and each pair is enclosed by a joint capsule filled with synovial fluid, so they are well lubricated.

Each thoracic vertebra has a pair of backward-facing facets on the upper side of its neural arch, rubbing against a pair of forward-facing facets on the lower side of the next vertebra above (FIG. 3.8a). Each lumbar vertebra has a pair of inward facing facets on its upper side, that rub against outward-facing facets on the next vertebra above (FIG. 3.8c). Successive vertebrae have their facets set at slightly

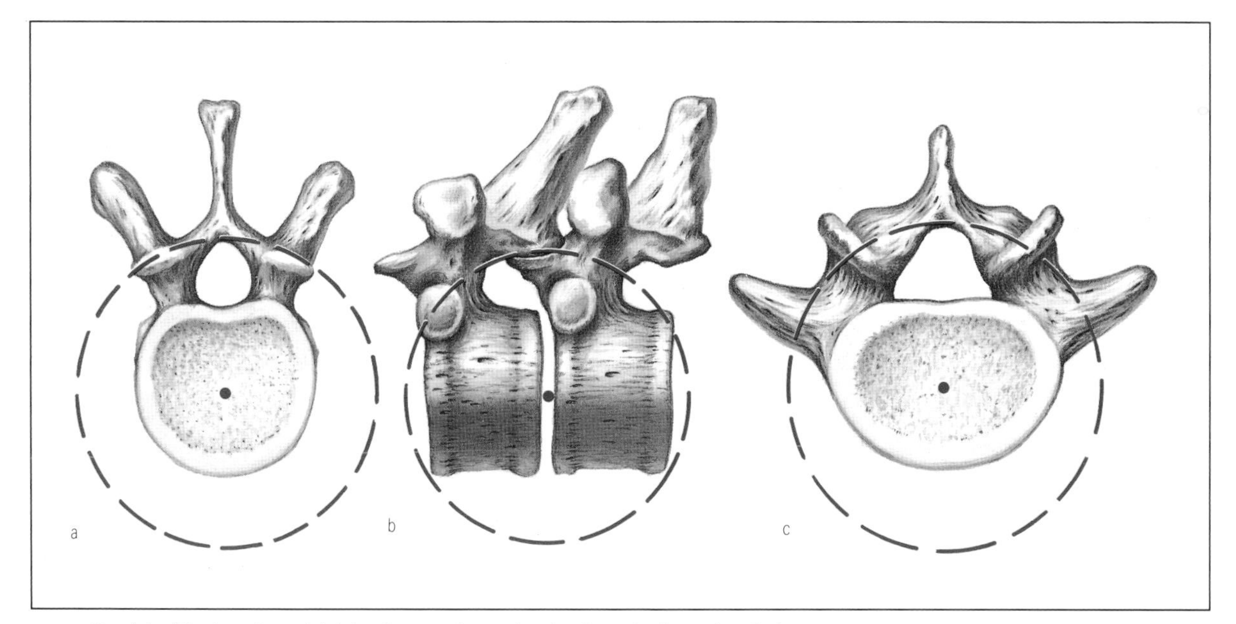

FIG. 3.8 *(a), (b) thoracic and (c) lumbar vertebrae, showing how the facets interlock.*

different angles, but here is a rough, simplified explanation of the general arrangement.

If successive vertebrae were attached to each other only by the intervertebral discs, they could move a little relative to each other by rotation about any axis through the centre of the disc: they would behave as if connected by ball-and-socket joints. In FIG. 3.8a,b the circles represent a sphere centred in the middle of a thoracic intervertebral disc. A ball and socket joint in that position would have no effect on the mobility of the joint but would allow the same movements as if the vertebrae were connected only by the disc. Now notice that the facets lie approximately on the surface of the sphere so that they are, in effect, small parts of the ball and socket that we have been imagining. They can have little effect on the mobility of the thoracic spine. However, the thoracic vertebrae are connected to ribs which in turn are connected to the sternum (breastbone), and so restrict movement. Each thoracic vertebra can bend forward and back relative to its neighbour through a range of only 4 to 6 degrees, but it can bend from side to side and twist about the long axis of the spine through ranges of 8 to 9 degrees.

The facets of the thoracic vertebrae have little effect on the mobility of the spine because they lie (approximately) on our imaginary sphere. The facets of the lumbar vertebrae, however, are approximately perpendicular to the surface of the sphere (FIG. 3.8c), and prevent relative rotation of the vertebrae about the long axis of the spine. Such movement between each vertebra and the next is restricted to a range of about 2 degrees, but 6 degrees side-to-side bending and 12 to 15 degrees forward and back bending is possible.

Because of these restrictions, different kinds of movement depend on different parts of the spine. When you turn your shoulders to left and right, keeping your hips stationary, most of the movement is between thoracic vertebrae. When you bend forward to touch your toes, most of the movement in the spine is between lumbar vertebrae, which allow the first 60 degrees or so of forward bending: to bend more, you tilt your pelvis forward.

As you bend forward you feel tension in your principal back muscles (anatomists call these muscles erector spinae). These lie on either side of the neural arches and run all the way from the base of the skull to the pelvic girdle. You use them to straighten your back after leaning forwards, especially if you are lifting something heavy. Though the muscle itself is so long it is built up of much shorter bundles of muscle fibres, many of which run short distances from one vertebra to another. The main part of the muscle attaches to the vertebrae of the lower back by way of a strong sheet of tendon.

Lifting strength has been measured by means of a force transducer with a handle, screwed to the floor. Healthy young men were asked to pull upwards on the handle as hard as they could and it was found that they could manage, on average, an upward pull of 900 newtons, about 1.3 times body weight. FIG. 3.9 shows that this pull required very large forces in their backs. Imagine the body sliced across just below the third lumbar vertebra and think of the forces on the upper part. The weight of this part (400 newtons) and the 900 newton load are both trying to bend the back forwards. Their moments about the joints between the lumbar vertebrae must be balanced by tension in the back muscles. The calculation works just like the shoulder calculation (FIG. 3.1) and like it gives a muscle force that is much larger than the load, because the muscle acts close to the joint. The calculated muscle force is a massive 7200 newtons (0.7 tonne force).

We did not calculate the reaction at the joint in the shoulder calculation, but in this case we will want to know it. The triangle of forces in FIG. 3.9 shows that it must be 7900 newtons (0.8 tonne force). This force compresses the vertebrae of the lower back.

The accuracy of calculations like this has been checked by experiment. Forces compressing the vertebrae also compress the intervertebral discs, raising the pressure in their pulp. This pressure has been measured in brave volunteers through a hypodermic needle stuck into a disc, while they lifted weights and performed other exercises. The relationship between force and pressure was checked by experiments on groups of vertebrae taken from

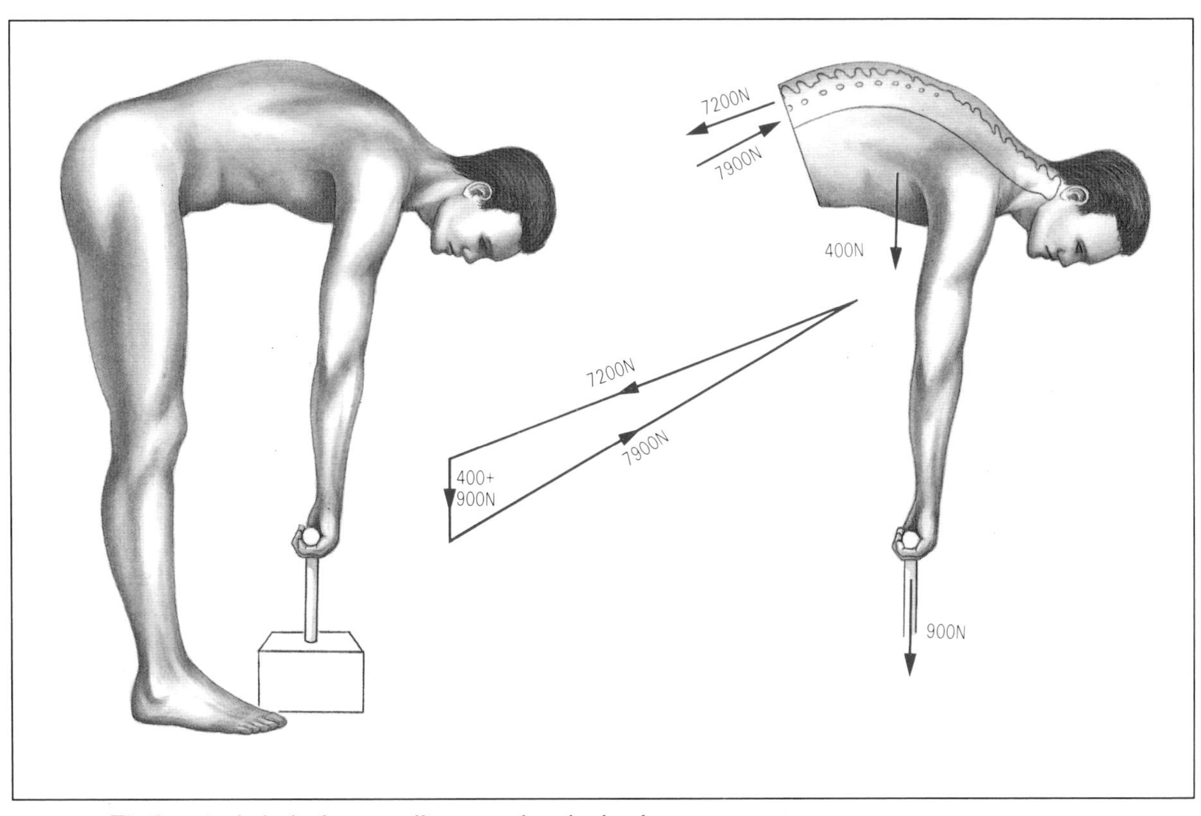

FIG. 3.9 *The forces in the back of a man pulling upwards as hard as he can.*

cadavers. These were squeezed with known forces in a machine like those used for stretching tendons (FIG. 3.4; the machines can squeeze or pull as required) while the pressure in the intervertebral discs was measured. These experiments confirmed that the forces that had been calculated were reasonably accurate.

The huge forces that act in heavy lifting make one wonder about the strength of the vertebral bodies. Many vertebrae have been squeezed in testing machines to measure the forces needed to crush them. For obvious reasons, most of the tests were made on vertebrae from elderly people who had probably been inactive for some time before they died. Many of these vertebrae were found to be far too weak to stand the forces of around 8000 newtons that act when young men lift heavy weights. However, one set of tests used lumbar vertebrae only from men aged 22 to 46 years, who had been mobile immediately before death. The average strength of these vertebrae was 10 000 newtons.

Many people injure their backs by lifting heavy loads. Nurses (who lift patients) and construction workers are particularly at risk. Backs get hurt in a large proportion of injuries at work. We must distinguish between accidents from things falling on people, from people getting caught in machinery and suchlike, and accidents due to over-exertion. Almost half of the over-exertion accidents reported to the Factory Inspectorate in Britain in 1979 were back injuries due to lifting.

Many back injuries are muscle tears but some involve damage to an intervertebral disc. The pulp may get squeezed to one edge, making the disc bulge out and causing low back pain. In more severe cases the pulp may burst out through the edge of the disc and press on one of the nerves emerging from the spinal cord. This happens most often to the disc immediately above the sacrum or the one above that. The nerves that these discs may press on run down the leg and the patient may feel sharp pain in his leg although the cause of the trouble is in his back.

A layer of muscle forms a wall round the abdomen,

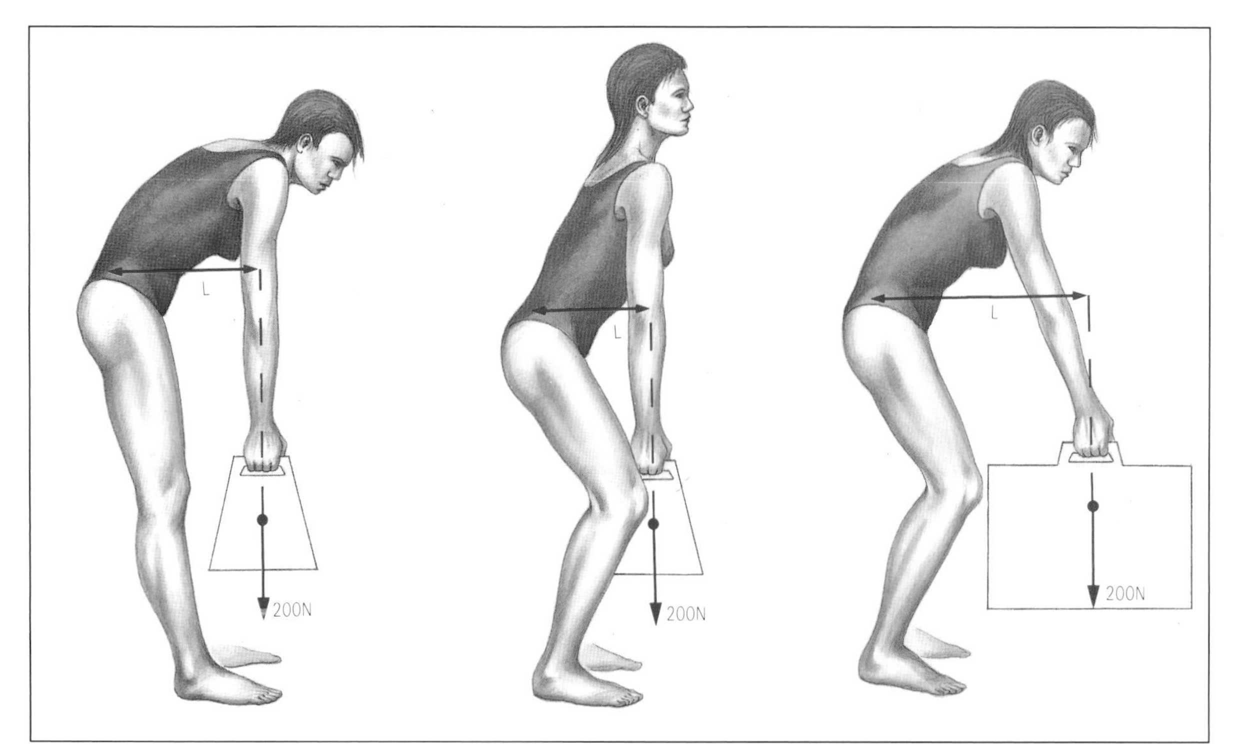

FIG. 3.10 *Three ways of lifting a weight. After M. Lindh (1989) Biomechanics of the lumbar spine, in M. Nordin & V.H. Frankel (Eds)* Basic biomechanics of the musculoskeletal system, *183–207. Lea & Febiger, Philadelphia.*

a FIG. 3.11 *Stages of the clean and jerk. © Colorsport.* b c

enclosing the intestines and other guts. It is often tensed when heavy objects are lifted, increasing the pressure in the abdomen. It has been claimed that this pressure may balance a useful fraction of the load, reducing the tension needed in erector spinae and the compressive load on the spine. In an experiment the subjects swallowed a pressure-sensitive radio transmitter, which may sound like a large mouthful but was actually a small cylinder, only 25 millimetres long and 9 millimetres in diameter. This passed down to the abdominal guts, and while there transmitted a signal giving the pressure in the abdomen. Large pressures were often recorded at the start of a lift, but when a weight was held at constant height for a while, there was very little pressure difference between the abdomen and the air outside.

People who lift heavy weights need to know the best way to avoid injuring their backs. FIG. 3.10 shows a woman lifting a weight in three different ways. In (a) she is bending over with her legs straight but in (b) she is bending her legs and keeping her back straight. The moment arm of the load about the lower back (the distance L) is smaller in (b) so the bending moment on the back is less; less force is therefore needed in the erector spinae and there is less load on the vertebrae. This would not be true if the object were too large to fit between her knees and had to be lifted from further forward (c). In that case L would be larger and bent-legged lifting would impose larger loads on the spine than straight-legged lifting.

Similar arguments tell us that it is less hard on the back to lift a 20 kilogramme block of lead, than a 20 kilogramme box of cotton wool. The lead is smaller so its centre of gravity can be kept closer to the body and its weight does not exert such large moments about the lower back.

FIG. 3.11 shows a weightlifter performing one of the standard competition lifts, the clean and jerk. In the clean he lifts the barbell to shoulder level and in the jerk he raises it to arms' length above his head. Good lifters can lift very heavy weights: the world middleweight record (for contestants of up to 75 kilogramme body mass) is 215.5 kilogrammes.

FIG. 3.12 shows the forces exerted on the floor by a foot of a 104 kilogramme (1020 newton) man lifting a 165 kilogramme (1610 newton) weight. You might expect to see the force plate registering $\frac{1}{2}$(1020 +

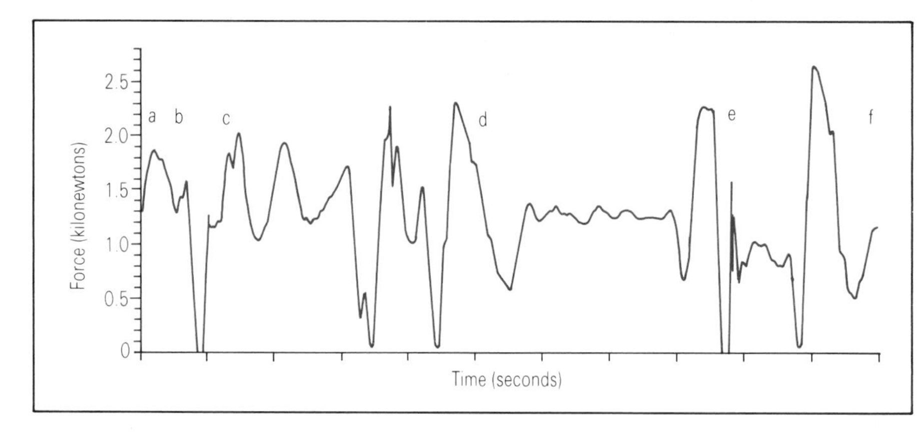

FIG. 3.12 *A record of the vertical component of the force on one foot, during weightlifting. The letters refer to the stages in* FIG. 3.11. *Re-drawn from M.W. Whittle, A.J. Sargeant and L. Johns (1988) in G. de Groot* et al *(Eds)* Biomechanics *XIB, 885–888. Free University Press, Amsterdam.*

1610) = 1315 newtons (the factor of one half is there because only one foot was on the plate), but the force fluctuates dramatically. Notice particularly that there are times when the plate, with man and weight on it, registers no force.

To understand this we need to realize that the technique of weightlifting is designed to overcome the relative weakness of our arms. Our arm muscles cannot exert as large moments at the arm joints as our leg muscles can at the leg joints. In all the positions shown in FIG. 3.11, the elbows and shoulders are almost directly above or below the bar, so the weight exerts relatively small moments about them. However, to get from position (b) to position (d) the lifter must pass through an intermediate position with forearms horizontal, a position that cannot be held because excessive moments would act at the elbow. The solution is to start the weight moving upwards and then suddenly to bend the knees, letting the body drop from position (b) to position (c). During this movement the body is falling freely and the feet exert no force on the ground. Straightening the knees again brings the lifter to position (d). Similarly, to get from position (d) to (f) the body is allowed to drop to position (e), with the force on the feet again falling to zero, and the knees are straightened yet again. Notice that the complete lift involves straightening the knees three times. Most of the work of weightlifting is done by the leg muscles.

That account explains the two stages in the force record (FIG. 3.12) when the force falls to zero. Other fluctuations in the record are due to the weight and body being accelerated and decelerated, either as intentional parts of the movement or in unintended wobbling, or to the foot being moved.

Finally for this chapter I turn from weightlifters to adolescent girls, the commonest victims of the disease known as idiopathic scoliosis. The spine twists and buckles as shown in FIG. 3.13 giving them (in extreme cases) a hump-backed appearance. Once it has started buckling, compression loads on the spine can more easily make it buckle more, so surgery may be needed not only to correct it, but to prevent it from getting worse.

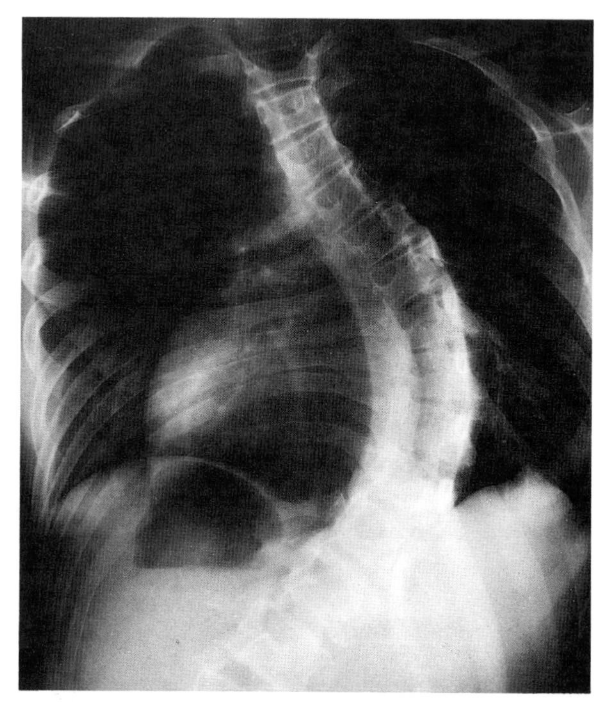

FIG. 3.13 *An X-ray picture of the spine of a girl suffering from idiopathic scoliosis, supplied by Professor R.A. Dickson of the Department of Orthopaedic Surgery, St. James' Hospital, University of Leeds.*

CHAPTER FOUR

STANDING

To understand standing, we need to know something of the structure of our legs. They attach to the bones of the pelvic girdle which are in the lower part of the trunk, rigidly attached to the sacral vertebrae (FIG. 4.1). The upper edges of the pelvic girdle form the bony ridges just below the waist, close under the skin of our hips. The hip joint is a ball and socket: the socket is in the pelvic girdle and the ball on the femur, the bone of the upper leg. The knee is a hinge between the femur and the tibia, which is the principal bone of the lower leg, but a much more slender bone (the fibula) lies alongside the tibia. The ankle is a universal joint: you can tilt your foot toe up or toe down, and side up or side down. Thus there are three degrees of freedom of movement at the hip, one at the knee and two at the ankle, a total of six between trunk and foot. This is the minimum needed for us to be able to place our feet where we want (within limits) on level or tilted surfaces. (This principle was explained in the discussion of arms in Chapter 1). However, the fibula cannot rotate on the tibia as the radius rotates on the ulna, so the leg lacks the extra degree of freedom that is found in the arm.

Hips are much less easily dislocated than shoulders because their sockets are much deeper (almost half a sphere). However, they are very apt to fail by osteoarthritis. Fortunately, hip replacement operations have been extremely successful since Sir John Charnley introduced his prosthesis in the 1960s. About 300 000 hip replacements are fitted annually, throughout the world. They enable large numbers of mostly elderly people, whose lives would otherwise be very restricted, to move around freely.

The Charnley hip replacement is a ball and socket joint consisting of a stainless steel ball that fits into a plastic (high density polyethylene) socket (FIG. 4.2). The ball is mounted on a curved steel shaft, which is pushed down the marrow cavity of the femur and cemented in place after the original femoral head has been cut off. The plastic socket is cemented into a hole cut in the pelvic girdle. These prostheses often give good service for many years but sometimes have to be replaced, most often if the cement in the femur works loose.

Osteoarthritis of the hip often has no obvious cause, other than normal wear and tear over many years. In other cases it seems to be due to the unusually severe loads on the joint that occur in some sports. One of the early recipients of a Charnley hip was the Marquess of Exeter who (as Lord Burghley) had won the 400 metre hurdles in the Olympic games of 1928 and had presumably damaged his hips by hurdling. He was so pleased with the mobility it gave him that when his first Charnley hip had to be replaced he mounted it instead of the usual emblem on the front of his Rolls-Royce.

The knee is a hinge like the corresponding joint in the arm between humerus and ulna, but is constructed very differently. Instead of surfaces that fit closely together it has rounded surfaces on the femur and much flatter ones on the tibia (FIG. 4.3). Collateral ligaments on either side hold the bones together and allow the knee to bend. You might think that these ligaments would be enough, but a ligament held by them alone would be wobbly because ligaments running at right angles to the contact surfaces would be unable to prevent small sliding movements of one bone on the other: similarly, if you want to moor your boat firmly you should not arrange your ropes at right angles to the quay, but at

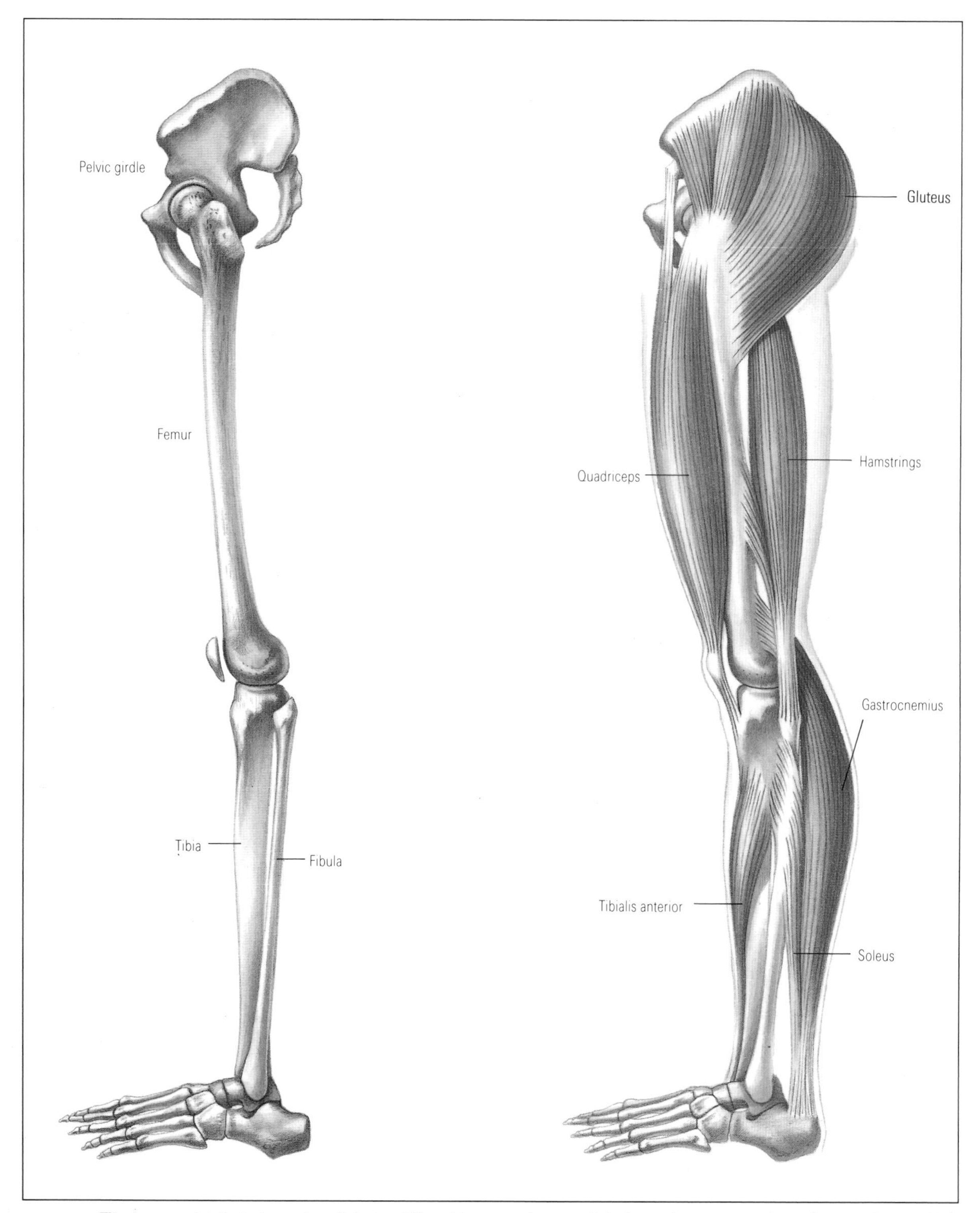

FIG. 4.1 *The bones and principal muscles of the leg. The adductors and some of the hamstrings are not shown because they would be completely hidden.*

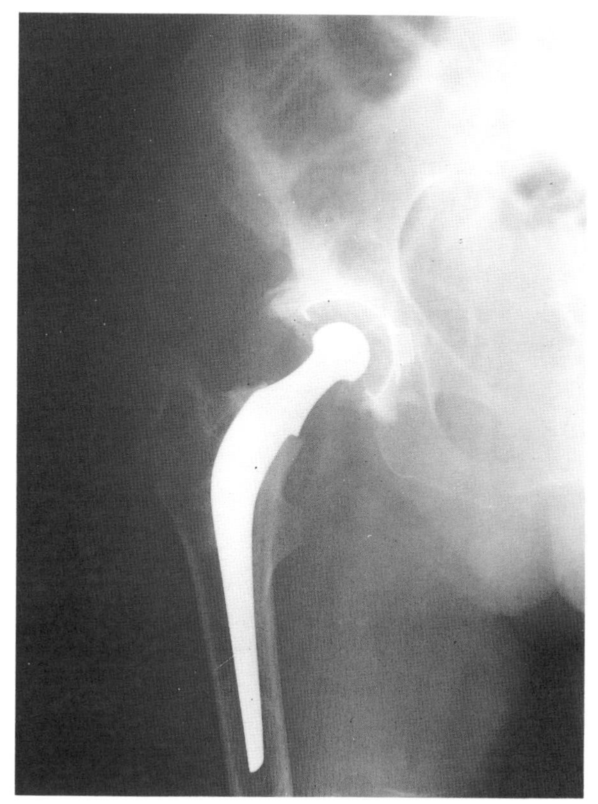

FIG. 4.2 *A Charnley replacement hip joint, seen in position in an X-ray picture supplied by Professor V. Wright, Department of Rheumatology, University of Leeds.*

a smaller angle. Angled ligaments will be most effective in holding the knee together firmly. At least two ligaments sloping in opposite directions are needed.

The knee end of the femur has two rounded surfaces with a groove between. In this groove, connecting the femur to the tibia, are the two cruciate ligaments, strong ligaments that cross over each other, one sloping forward and the other back. The angles between the ligaments change as the knee bends and extends (FIG. 4.3b,c).

FIG. 4.4 represents a model that shows how crossed, inextensible ligaments can allow a joint to bend and extend. However, for the ligaments to be kept taut, the shape of the joint surfaces must be precisely matched to the dimensions and arrangement of the ligaments. In the case of the model, the femur would have to be shaped as shown by the broken lines. Surprisingly, our knee joint surfaces do not seem to be ideally shaped. When our knees are straight the ligaments are taut and the joint behaves as a near-perfect hinge, allowing just one degree of freedom of movement. When they are bent the ligaments

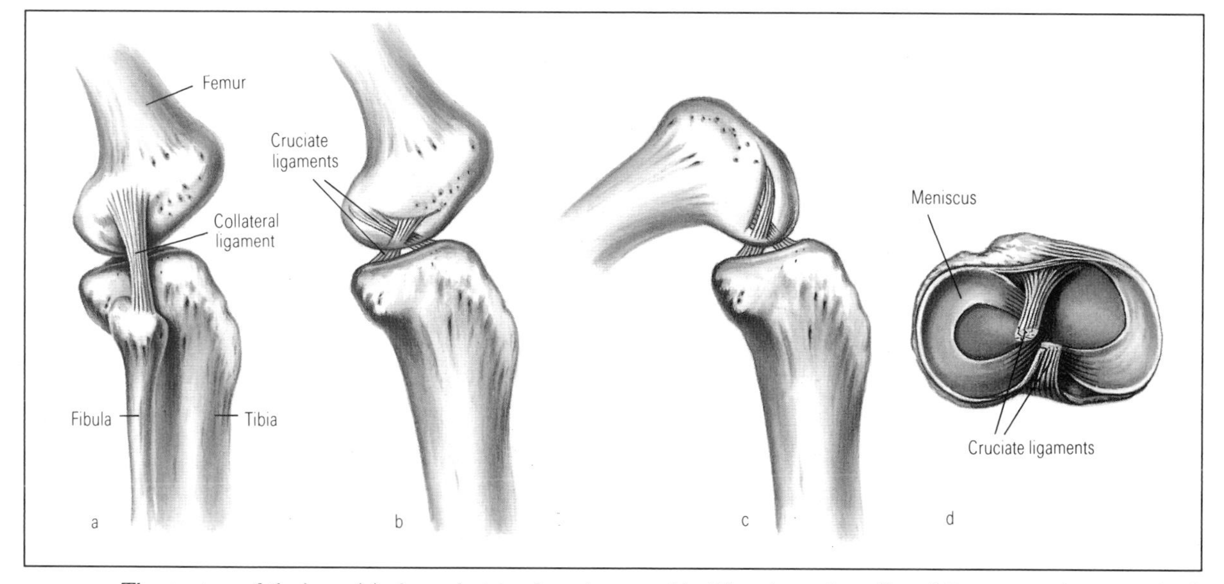

FIG. 4.3 *The structure of the knee. (a) shows the joint from the outer side. There is another collateral ligament on the inner side of the joint. (b), (c) show how the angles of the cruciate ligaments change as the joint moves. (d) shows the upper surface of the tibia with the femur removed but the menisci still in place. It also shows where the ligaments attach.*

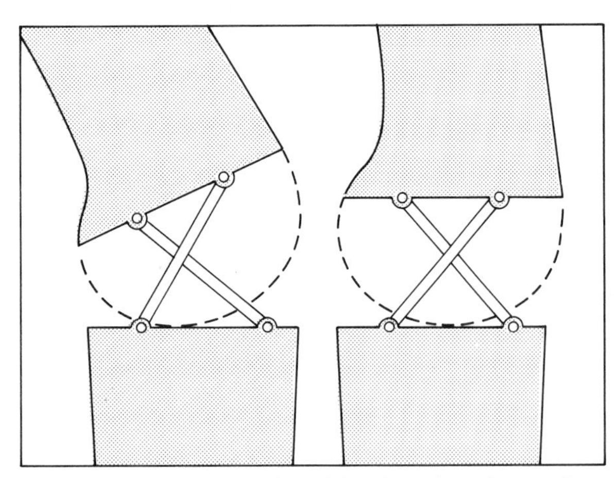

FIG. 4.4 *A model made of rigid bars hinged together, to show how the cruciate ligaments allow the knee to bend and extend. If the top of the 'tibia' were flat as shown the 'femur' would have to have the shape of the broken line, to allow the full range of movement of the model while maintaining contact between the two bones. A similar relationship would be needed between the shapes of a real femur and tibia, if the ligaments were to be kept taut throughout the range of movement.*

become slightly slack and the knee becomes a little wobbly, allowing small movements in other degrees of freedom. (This is difficult to demonstrate on yourself because you cannot get a firm grip on the femur). Experiments have shown that when the knee is bent at a right angle, with the femur rigidly clamped, very small forces are enough to move the tibia from side to side through an angular range of 8 degrees or to rotate it about its long axis through 24 degrees.

If the knee were held together only by the cruciate ligaments, they could be twisted around each other, tightening them up, by rotating the big toe inwards; and unwrapped, slackening them off, by rotating the toe outwards. The collateral ligaments prevent that from happening.

The gap between the rounded surfaces of the femur and the flatter surface of the tibia is largely filled by two crescent-shaped wedges of fibrous cartilage, the menisci. Each meniscus is held in place by ligaments. Without menisci, the femur would press on only a very small area of the tibia, setting up large stresses in the articular cartilage, but the menisci spread the load over a larger area. They frequently get torn in sport, apparently as the result of the knee being twisted while bent and heavily loaded. A torn meniscus can be removed surgically without any obvious ill effects, but people whose menisci have been removed seem more likely to suffer later from osteoarthritis of the knee.

One possible advantage of having the knee built as it is, instead of with close-fitting bones like the elbow, is that the crossed-ligament arrangement makes the femur move backward on the tibia as the knee bends and forward as it extends (FIG. 4.3b,c). The parts of the joint surfaces that take the load, both on the femur and on the tibia, change as the

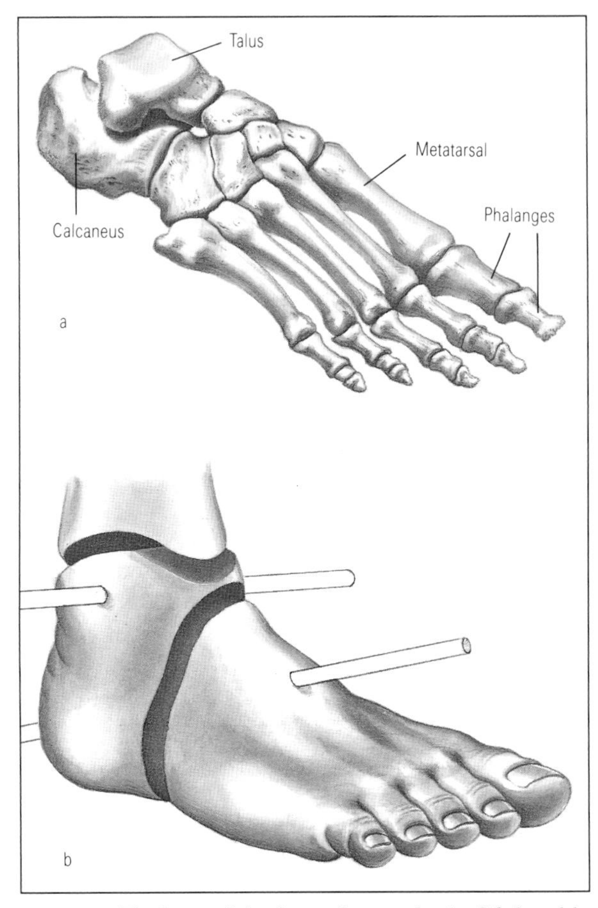

FIG. 4.5 *The bones of the foot and a greatly simplified model to show how the ankle joint works.*

joint moves. If cartilage in a loaded part of the joint gets the lubricating fluid squeezed out of it, it will be unloaded and get time to soak up fluid again when we change position.

Ankle joints are bafflingly complicated: there are a lot of small bones (as in the wrist) that move relative to each other. Two of these bones deserve special mention: the talus that forms the joint with the lower end of the tibia and the calcaneus that forms the heel and runs forward under the talus. When you bend your ankle toe up and toe down, most of the movement is between the tibia and the talus. When you rock it from side to side, tilting the sole of the foot inwards and outwards, most of the movement is between the talus and the other ankle bones. You will get some feeling for the working of the ankle if you simplify and think of it as two hinge joints: a hinge with a transverse axis between tibia and talus and another with its axis running along the length of the foot, between the talus and the other bones (FIG. 4.5). Two hinge joints as in a universal joint (FIG. 1.2b) allow two degrees of freedom of movement. I do not fully understand the complicated array of ligaments that hold the ankle bones together (and I do not think anyone else does, either) so I will not try to explain them.

The principal muscles of the leg are shown in FIG. 4.1, as well as the bones. The gluteus muscles are the flesh of the buttocks. They connect the upper end of the femur (which you can feel as a bony lump at the hip) to the pelvic girdle and the lower end of the vertebral column. The main action of their superficial part is to make the leg swing back at the hip joint, but the deeper parts of the muscle run at a different angle and swing the leg out to the side (FIG. 4.7).

The big muscles of the thigh are the hamstrings and adductors (behind) and the quadriceps group (in front). The hamstring muscles attach to the lower end of the pelvic girdle close to the bony lump where you get uncomfortable if you sit for long on a hard chair. At their other end they attach to the femur and tibia, near the knee: you can feel their tendons behind and to either side of the knee, when the knee is bent. They tend to swing the leg back at the hip (like the gluteus) and also to bend the knee. The adductor muscles lie nearer the inner face of the thigh. They attach to the pelvic girdle in the crotch and (at their other end) to the femur. When they shorten, they pull the leg in towards the mid line. The quadriceps muscles attach at their upper end mainly to the femur (but one part, the rectus, attaches to the pelvic girdle). At their lower end they attach to the tibia by means of a thick tendon that has the kneecap embedded in it. When they contract they straighten the knee but the rectus also tends to swing the leg forward at the hip. Because their tendon runs round the front of the bent knee it presses on the end of the femur. Similarly the direction of pull of a rope can be changed by running it over a pulley. A pulley rotates as the rope moves but the femur of course does not: the kneecap slides over the femur as the knee bends and extends, and the surfaces must be lubricated to reduce friction and wear. The back of the kneecap and the groove in the femur in which it slides are covered by cartilage and bathed in synovial fluid, so they are lubricated like the other joints of the limbs.

The kneecap is not merely a device to allow the quadriceps tendon to slide over the femur like a rope on a pulley. The force in a rope is the same on both sides of a pulley but the quadriceps tendon pulls less hard on the tibia than the muscle pulls on it. This has been demonstrated by experiments on knees, like the one shown in FIG. 4.6a. Force transducers were attached to the tendon above and below the kneecap. They were pulled on in appropriate directions, with the kneecap in the positions it would occupy at various knee angles. The force registered by transducer B equalled the force on transducer A when the knee was straight and the two tendons in line with each other, but when the knee was bent to a right angle the force on B was only half that on A.

This reduction of force by the kneecap may seem like a disadvantage, but it is accompanied by magnification of movement. In the pulley system shown in FIG. 4.6b every centimetre movement of the hand towards the left results in one centimetre lifting of the weight. In the device shown in FIG. 4.6c, which works on the same principle as the knee, one centi-

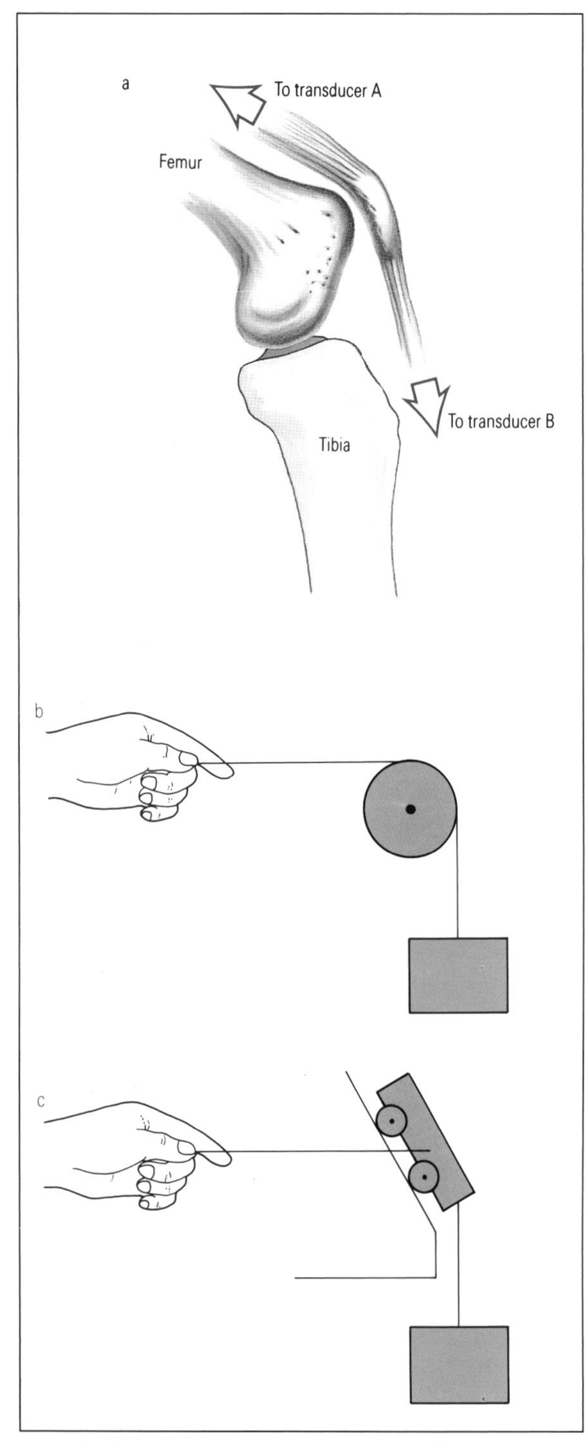

FIG. 4.6 *Diagrams to show how the kneecap affects the movement of the knee. (a) shows an experiment with two force transducers. (b) shows a rope running over a pulley and (c) is a model that magnifies movement by the same principle as the kneecap.*

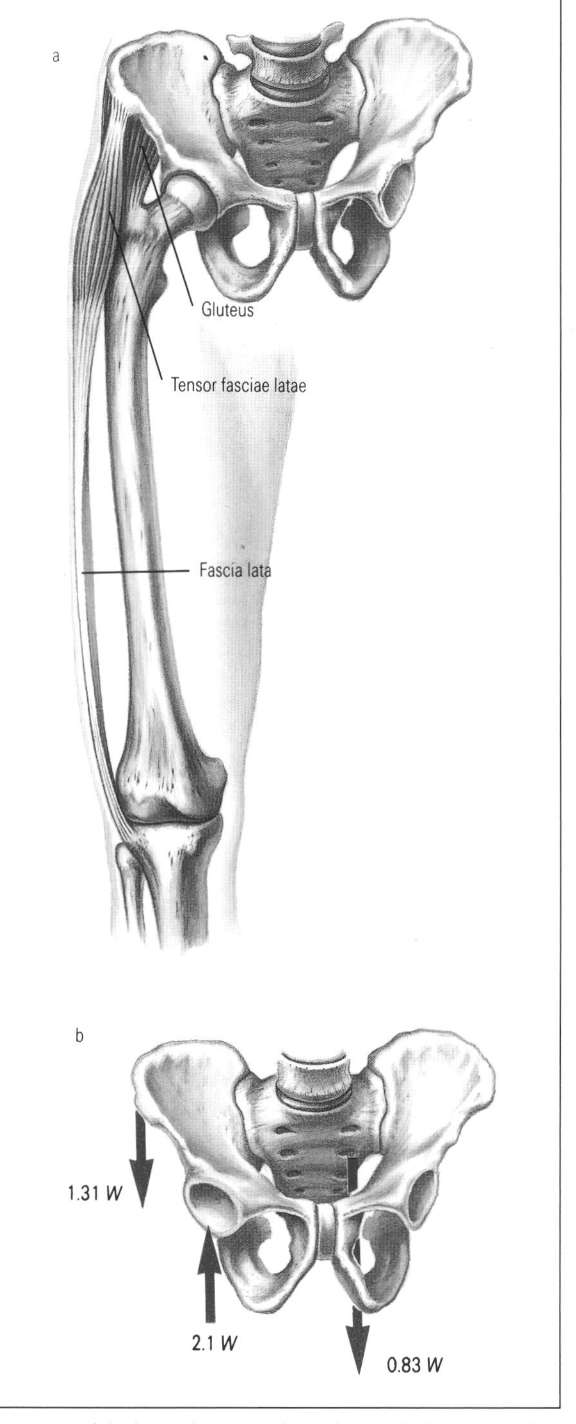

FIG. 4.7 *(a) shows the tensor fascia lata and gluteus muscles in place on the skeleton and (b) shows how the forces in them and in the hip joint can be calculated for someone standing on one leg.*

metre movement of the hand results in about two centimetres movement of the weight, but the force exerted by the hand has to be about twice the weight because the work (force multiplied by distance) done on the weight cannot be more than the work done by the hand. Similarly the kneecap mechanism gives more and faster movement of the joint for a given shortening of the muscles, than a similarly-placed pulley would give.

The tensor fasciae latae is a muscle that attaches to the upper edge of the pelvic girdle, just below the waist (FIG. 4.7). Its tendon is not a round cord, like most other tendons, but a broad sheet. This sheet, the fascia lata, runs down the outer side of the thigh, close under the skin, and attaches to the outer surface of the knee. Shortening of tensor fasciae latae swings the leg out to the side.

The gastrocnemius and soleus muscles make up most of the flesh of the calf. Both attach to the calcaneus (heel bone) through the Achilles tendon, the thick tendon that you can feel through your skin behind the ankle. When they shorten they pull the heel up, making the toe point down. The soleus attaches to the backs of the tibia and fibula but the gastrocnemius attaches to the femur, close behind the knee, and so tends to bend the knee as well as to move the ankle. The tibialis anterior muscle runs down the front of the tibia, connecting it to the foot bones close in front of the ankle. It tends to bend the ankle, making the toes point up.

There are other muscles in the leg but these are the largest and are also the ones that I will want to mention again, later in this book.

Now that the bones and muscles of the legs have been described I will explain how we stand on them. The first principle of standing is that your body's centre of gravity must be over one of the feet, or over a point between the feet: otherwise you would inevitably topple over. The position of the centre of gravity relative to the feet can be discovered by having a person stand on a force plate that registers not only the force that acts on it, but also the position of the centre of pressure. This needs some explanation. The forces that our feet exert on the ground are actually distributed over most of the area

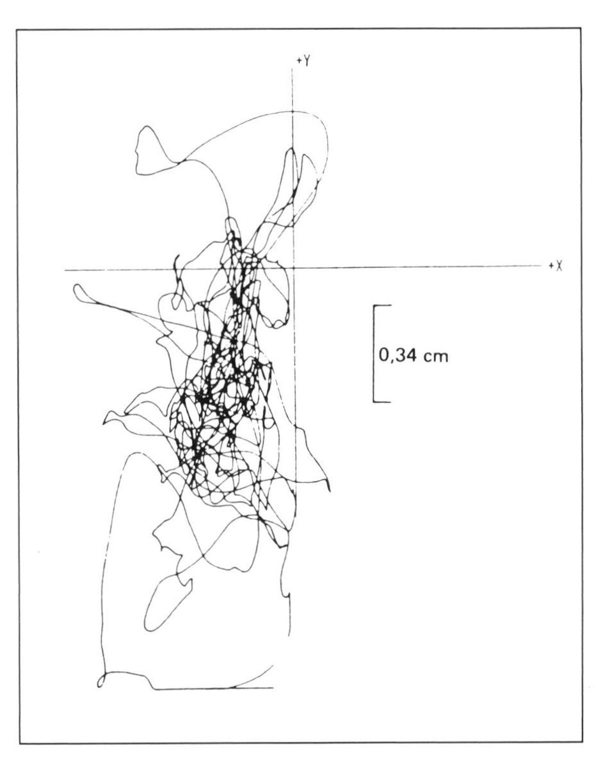

FIG. 4.8 *The track of the centre of pressure of a healthy child standing on a force plate, during 30 seconds. From G.F. Harris* et al *(1982)* J. Biomechan. **15**: *741–745*.

of their soles. However, for purposes of calculation we can replace these distributed forces by a single force acting at one point, the centre of pressure. Similarly, gravity acts on all parts of our bodies but for many purposes we can think of our weight as a single force acting at the centre of gravity. Force plate records of people standing show the centre of pressure moving around a little (we do not stay quite still) but it usually stays within a circle of about 25 millimetres diameter between the feet (FIG. 4.8), or a little more for elderly people.

Ordinary force plates locate the centre of pressure by having force transducers at each corner. If the centre of pressure is near the rear of the platform, the transducers at the rear register most force and if it is near the right hand edge, most of the force is registered by the transducers at the right. Other instruments can be used to show how the force

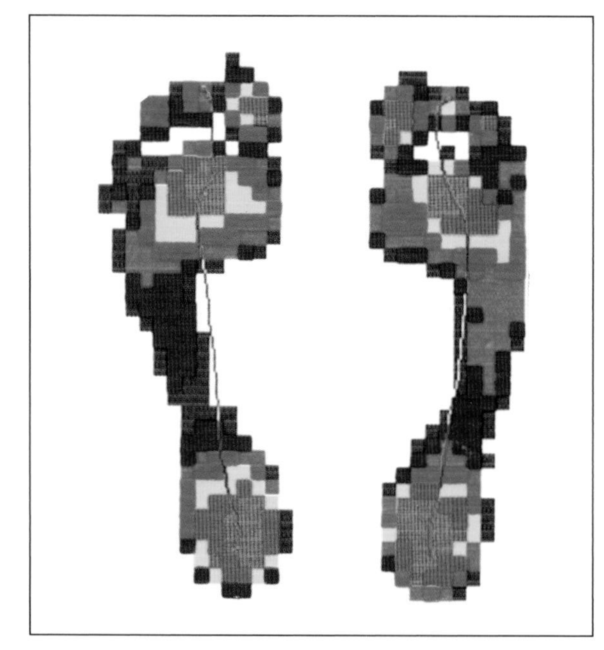

FIG. 4.9 *The distribution of pressure on the bare feet of a person standing on a pressure-sensitive plate. Gradations in pressure are shown by different colours, black indicating areas of low pressure and pink high.*

is distributed over the sole. One of these is the pedobarograph, whose prinicipal component is a glass plate covered by a thin sheet of plastic. The distribution of force when someone steps on the plate is discovered by an optical method that shows how firmly each part of the plastic sheet is pressed against the glass. The raw image is black and white but can be computer processed to give a colour display in which different colours represent different pressures, just as the colours in a contour map represent different heights. FIG. 4.9 shows that the force on the feet of a standing person is unevenly distributed, with the highest pressures under the heel and the metatarsophalangeal joints.

The soles of my (rather large) feet have a total area of about 350 square centimetres, so if my 700 newton weight were distributed evenly the pressure under them would be two newtons per square centimetre. Records like FIG. 4.8 show much higher pressures, up to about ten newtons per square centimetre, in some places, and lower ones in others.

The pedobarograph can record pressures for walking as well as for standing. Though the forces on the feet in walking are only twice as high as for standing the peak pressures are much higher because we have only the heel on the ground at one stage and only the ball of the foot at another. Peak pressures of about 60 newtons per square centimetre are usual, under the joint at the base of the big toe. Much higher pressures, up to four times as high, are recorded from some diabetic patients who develop ulcers on this part of the foot. A large part of their problem is that the layer of fatty tissue in the sole, that cushions the foot and spreads the load, is abnormally thin. The pressure peaks can be relieved by a rubber-like plastic insole.

When we stand in equilibrium, the centre of pressure on the feet must be vertically under the body's centre of gravity and the upward force on the feet must equal body weight. The centre of pressure is in front of our ankles so we have to use our calf muscles to prevent ourselves from falling over forwards. Get a friend to stand comfortably and feel the Achilles tendon through the skin behind their ankle: you can feel that it is tight. The centre of pressure is also in front of our knee joints so our weight would tend to make us fall over forwards at the knees, if they could bend that way. The knees cannot bend forward, so there is no need for our knee muscles to be active when we stand. However, if we stand with our knees bent so that they are in front of the centre of pressure, we have to use our quadriceps muscles to prevent ourselves from collapsing backwards at the knee (FIG. 4.10). Now feel your friend's kneecap, pushing it gently from side to side. You will find that in normal straight-legged standing it moves freely from side to side, showing that the quadriceps tendon (in which it is embedded) is slack. In standing on bent knees it is held firmly in place in its groove in the femur by the taut tendon, and will not move easily.

These conclusions about muscle activity can be checked by recording the electrical activity in the muscles. In normal standing, the calf muscles (gastrocnemius and soleus) are active, but there is little activity in the muscles of the thigh. Indeed, there may be periods when no electrical activity is recorded either in the quadriceps muscles (at the front of the thigh) or the hamstrings (at the back).

Some experimenters have found activity mainly in the soleus muscle in human standing, but others have found quite a lot of activity in the gastrocnemius as well. The situation is much clearer in cats. When they stand quietly, the soleus is fully active but there is very little activity in the gastrocnemius. Why should the soleus be preferred?

The forces required for standing are much smaller than the maximum that the muscles can exert: we will find much larger forces acting on the feet when we discuss running and jumping in later chapters. That means that only some of the fibres of the gastrocnemius and soleus need be activated: which is it best to use? The soleus muscle crosses only the ankle joint but the gastrocnemius crosses ankle and knee and exerts moments about both. In standing people, moments are needed only at the ankle, so the soleus can do the job. However, there is also another reason for preferring it.

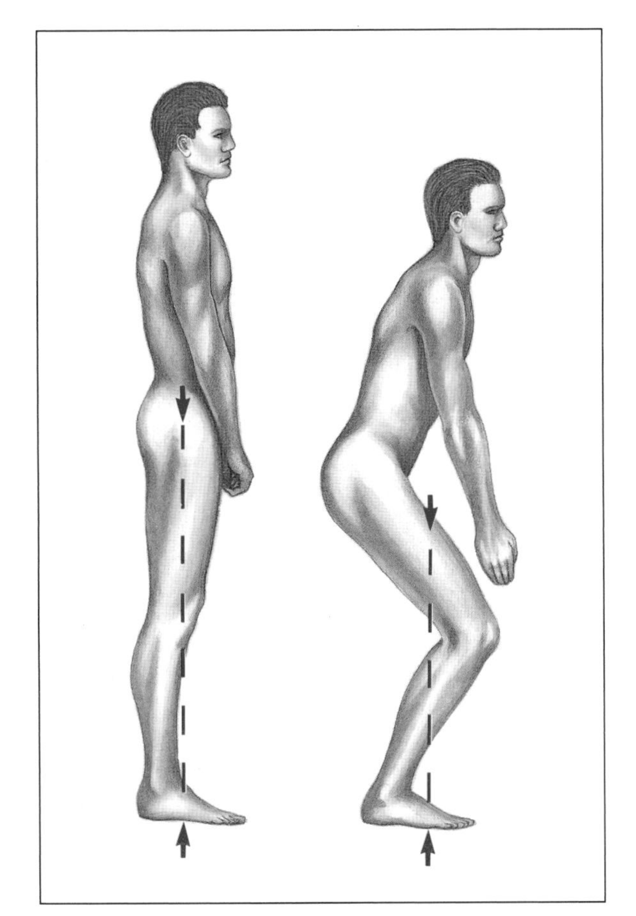

FIG. 4.10 *Diagrams showing the weight of the body and the force on the feet of a person standing with straight and bent legs.*

FIG. 3.3 shows that the rate of shortening of a muscle is limited: even when no external force resists its shortening, it cannot shorten above a certain rate. Different muscle fibres can shorten at different maximum rates. The ones that can shorten fast are needed for fast movements but the slow ones can maintain tension more economically, when speed is not required. Fast fibres are fast because their cross bridges detach and re-attach very rapidly. Remember that they work like people pulling on a rope hand over hand. If our hands make more pulls per second, we can pull a rope in faster. However, when a muscle is holding on, exerting force without shortening, the cross bridges do not remain attached but keep detaching and having to be re-formed, needing a new dose of energy each time. The cross bridges in fast muscle detach and re-attach more often, so needing more power (energy per unit time) to maintain the tension.

This has been demonstrated in experiments with rat muscles. Experiments like the one shown in FIG. 3.3 showed that toe extensor muscles (which consist almost entirely of fast fibres) can contract twice as fast as soleus muscles (whose fibres are almost all slow). In other experiments, the muscles were stimulated electrically but prevented from shortening, and their rates of oxygen consumption were measured. The toe extensor (the fast muscle) used oxygen twice as fast as the soleus, while exerting the same force. We can save energy by using only slow muscle fibres while standing still, but we have to activate the fast ones when we want to move fast. Obviously, we need all our muscle fibres to exert maximum forces, but if we are standing still and need only small forces, the slow fibres can supply them most economically.

FIG. 4.11 *Dancing* sur les pointes. *Photograph by Houston Rogers. By courtesy of the Trustees of the Theatre Museum, Victoria and Albert Museum.*

In cats, the soleus muscle consists entirely of slow fibres, and is used while the animal is standing still. The gastrocnemius consists mainly of fast fibres, some of which are recruited when the animal walks and more when it runs or jumps.

Standing on two legs needs remarkably little muscle activity but standing on just one requires the tensor fascia lata and gluteus muscles (FIG. 4.7) to hold the hip joint firm. Think of the forces that act on the part of the body that the leg supports (i.e. on everything except that leg). Each of a person's legs is about 17% of their total body weight (W) so the weight to be supported is $0.83W$. This acts at the centre of gravity of the supported part of the body, about 95 millimetres from the centre of the hip joint. The force of tensor fascia lata and gluteus acts along a line about 60 millimetres from the centre of the hip, so by the principle of levers this force must be $(95/60)\ 0.83\ W = 1.31\ W$. The upward force acting through the hip joint must balance these two downward forces, so it must be $(0.83 + 1.31)\ W$: the force at the hip joint must be 2.1 times body weight when we stand on one leg. Similar forces act in walking, when the body is supported first by one leg and then by the other. Much larger forces (six or more times body weight) act on the hip joint when we run and even larger forces in athletic jumping events. Perhaps we should not be surprised that the hip joint is particularly liable to osteoarthritis.

Female ballet dancers often stand *sur les pointes*, on the extreme tips of their toes (FIG. 4.11). FIG. 4.12 is from an X-ray picture of a dancer doing this. (The nails that seem to float in mid air are in the sole of her shoe.) Her whole weight was supported on this foot, when the X-ray was taken, and she was in balance. Therefore, a force equal to her weight must have been acting vertically upwards, on the tips of her longest toes (presumably mainly on the big toe). Notice how the joints in the big toe and in the ankle are stacked one above the other, close to the line of the force, so the moments about the joints must have been small. Standing on points is a feat of balance, rather than of strength. However, it does cause trouble, especially in the big toe, which tends to bend towards the others. This is called hallux

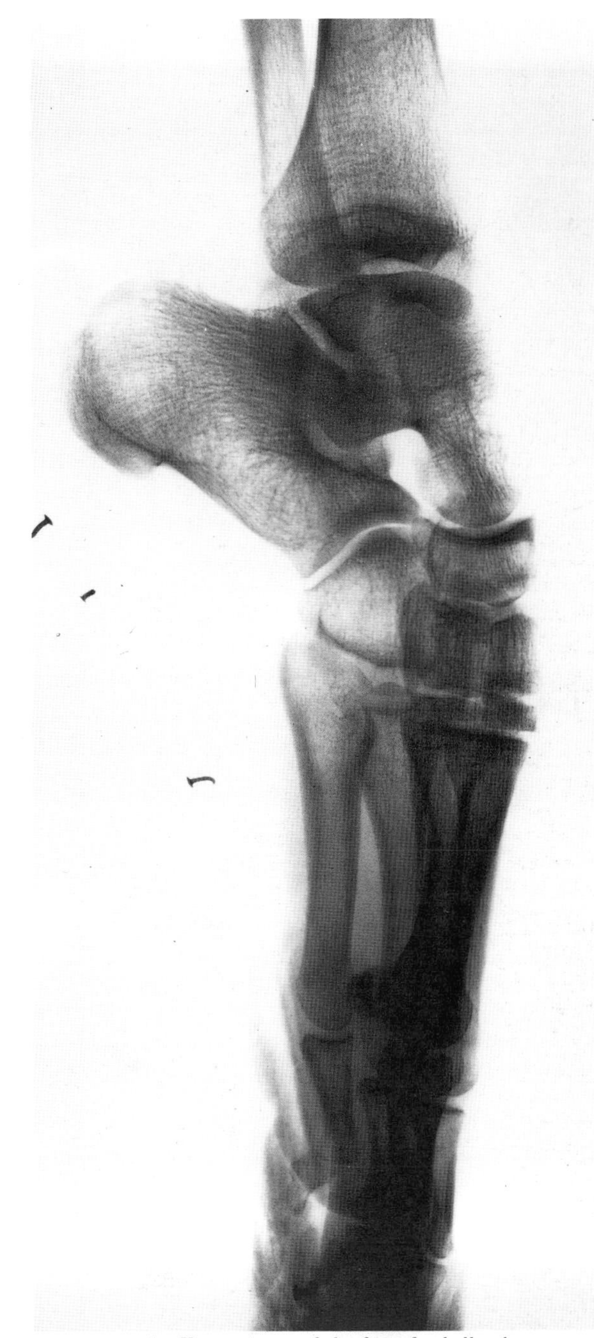

FIG. 4.12 *An X-ray picture of the foot of a ballet dancer standing on point, supplied by Mrs E. Whitmore, Department of Anatomy, Charing Cross and Westminster Medical School, London.*

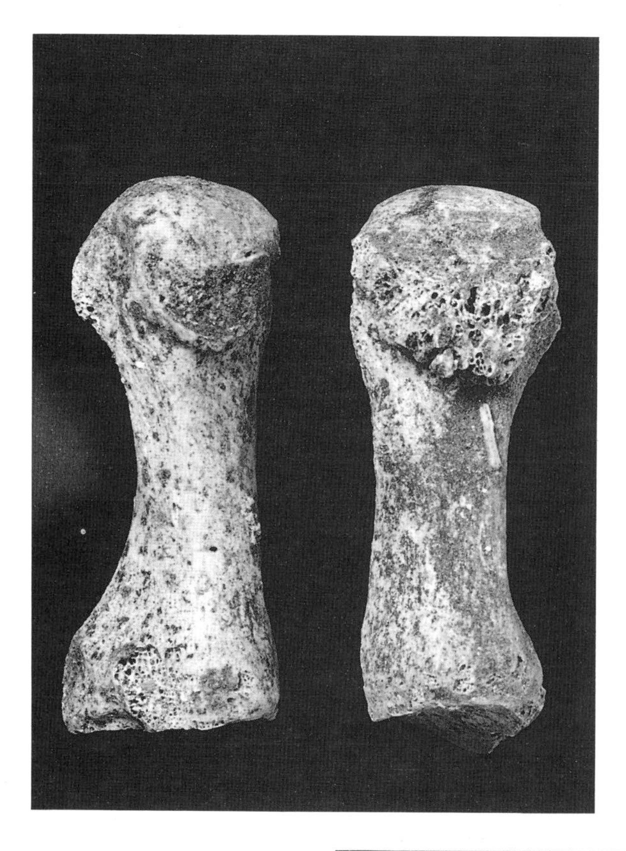

valgus, and one of its consequences is the development of a bunion, a bony outgrowth at the base of the toe. Hallux valgus and bunions are also common in non-dancers who cram their feet into tight-fitting shoes. Another common complaint of ballet dancers that is presumably caused by dancing is osteoarthritis of the joint at the base of the big toe.

Some 8000 year old skeletons found by archaeologists in Syria show severe osteoarthritis of the big toe (FIG. 4.13), and we must surely look for an explanation other than ballet dancing. One very plausible suggestion is that it may be the result of grinding grain in the manner shown in FIG. 4.14. The grain is on a slab of rock (a saddle quern) and is being rubbed with a smaller stone. Saddle querns of the same date as the skeletons have been found in the same area. Notice how the toes of the woman's right foot have been bent uncomfortably far forward to give the purchase she needs on the ground to push the rubbing stone forwards. Notice also how she has folded the left foot over the right as if to ease pain in it. Osteoarthritis seems as likely to result from prolonged repetitive work like this, as from the prolonged repetitive work in modern textile factories that was discussed in Chapter 3.

FIG. 4.13 *The metatarsal bones of the big toes of a Neolithic skeleton from Syria. The right metatarsal shows severe osteoarthritis, but the left is more normal. From T. Molleson (1989)* Antiquity **63**, *356–362.*

FIG. 4.14 *An ancient Egyptian figurine of a woman using a saddle quern. From J.H. Breasted (1948)* Egyptian Servant Statues. *Pantheon Books, New York.*

CHAPTER FIVE

WALKING

The main thing that our legs have to do when we walk is to support our weight, which requires a vertical force on the ground. Once we have got started on level ground there is very little need for any horizontal force to keep us going at a steady speed, merely enough to overcome air resistance. That argument suggests that we should walk like the man in FIG. 5.1, keeping the force on the ground almost precisely vertical and equal to body weight, throughout the stride. We do not walk like that except perhaps on ice whose slipperiness prevents us from exerting much horizontal force.

Walking on ice is tiring and FIG. 5.1 makes it easy to see why. At stage (a) the vertical force acts in front of the hip so the hamstring and gluteus muscles must be active, to balance the moment of this force about the hip. At stage (c) there is a large moment about the hip again and muscles such as the rectus, in front of the hip, must be active. Between these stages the body moves forward horizontally (if the ground force is always precisely equal to body weight) so the knee must be bent at stage (b). That is likely to result in large moments at the knee joint, requiring the quadriceps muscles to be active. Considerable tension would be needed in one muscle or another, at most stages of the stride. Muscle tension costs metabolic energy, so we would burn a lot of fuel walking like this. Also, we would rock uncomfortably back and forward: the force on the foot acting in front of the centre of gravity in (a) would tend to rotate the body anticlockwise and the force in (c) would tend to rotate it clockwise.

We do not walk like that. While each foot is on the ground we keep its knee almost straight. The body is higher when supported by a straight vertical leg (FIG. 5.2b) than when supported by straight sloping legs (FIG. 5.2d), so we rise and fall. If you watch someone walk in front of a wall marked by horizontal lines you will see the top of the head rise and fall about 35 millimetres, in every step. This implies that the force on the ground is not always equal to body weight, but fluctuates a little. Force plate records of walking show that this is true (FIG. 5.3). The average force on the ground over a complete stride must equal body weight but the force is greater at some stages and less at others.

Your body is rising between stages (a) and (b) of a stride (FIGS. 5.2 and 5.3) and falling between stages (b) and (c). Therefore at stage (b) you are accelerating downwards: at this stage the force on the supporting foot must be less than your weight. After stages (c) and (d) you are rising again so at stage (d) you are accelerating upwards: the total force on your two feet must then be greater than body weight. FIG. 5.3

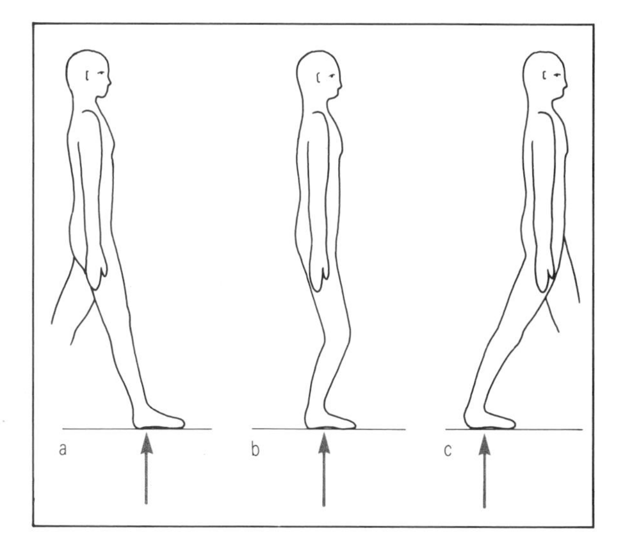

FIG. 5.1 *We do not walk like this.*

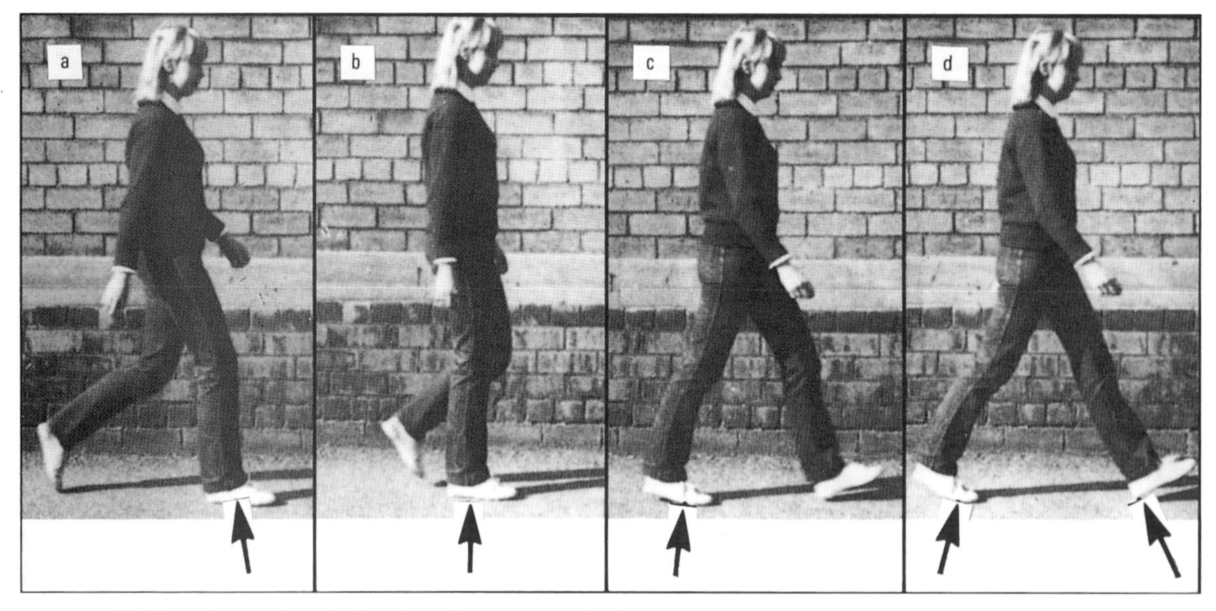

FIG. 5.2 *Walking. Arrows show the directions of the forces on the feet.*

shows how the left and right feet exert patterns of force that add up to more than body weight at stage (d) of the stride, but less at stage (b). The faster you go, the less time there is to rise and fall: the accelerations have to be larger, and so do the fluctuations of vertical force.

Force plate records also show that the forces on the feet have horizontal components (FIG. 5.3). The average force is approximately zero but there is a forward force on the foot, accelerating the body, at one stage of each step, and a backward force, decelerating it, at another. Thus the resultant force

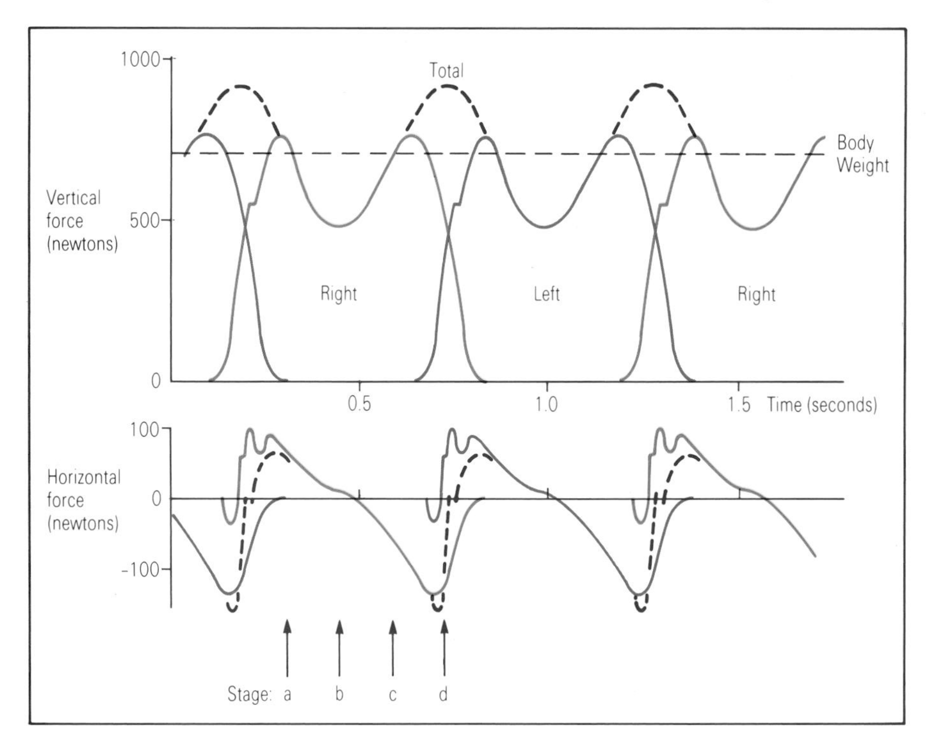

FIG. 5.3 *The vertical (above) and horizontal (below) components of force that a walker's feet exert on the ground. This graph has been built up from a force plate record of a single footfall, by repeating the record at appropriate intervals.*

on the foot acts upward and backward at one stage, upward and forward at another: the effect is that the force is always more or less in line with the leg. That way, the moments of the force about the knee and hip are always small and rather little of the muscle in the leg has to be activated. Also, since the centre of gravity of the body is only a little above the hips (about 50 millimetres above them) the forces on the feet keep fairly nearly in line with the centre of gravity and have little tendency to make the body rock.

FIG. 5.4 shows legs that remain perfectly straight while the foot is on the ground, making the body's centre of gravity move in arcs of circles. One foot is set down and the other lifted at the instant of changing from one arc to the other. At one instant the body is moving downwards and at the next it is moving upwards, so its velocity is changed instantaneously, requiring an infinite acceleration and an infinite force. Walking like that is impossible.

We actually round off the corners between the circular arcs, as shown by the blue line at (3) in FIG. 5.4. Thus the changes in velocity are not instantaneous, and infinite forces are not needed. Also, we do not lift one foot at the instant of setting down the other, but have both feet simultaneously on the ground for about one quarter of the time.

Following the blue line at (3) means that the centre of gravity is then above the circular arcs, so the legs must be longer then than at stages (2) and (4). However, the knee of the supporting leg is straight at these stages, and we want it straight to minimize the forces in our muscles. How then can the legs be longer at stage (3)? The answer is that we have big feet and do not keep them flat on the ground, but set our heels down first at the start of the step and rise up on our toes at the end. This raises the centre of gravity a little higher than it would otherwise be, at the stage of the step when both feet are on the ground (FIG. 5.5). The heel-and-toe style of walking enables us to keep our knees straight, while the feet are on the ground.

In describing how we walk I have emphasized the straightness of our legs, explaining that this keeps the forces needed in muscles small, and so saves energy. In many other books you will find a different explanation of the advantage of the human style of walking, which goes like this. Mechanical energy can take various forms including potential energy (energy due to height) and kinetic energy (energy due to movement). If you drop a stone it falls and so

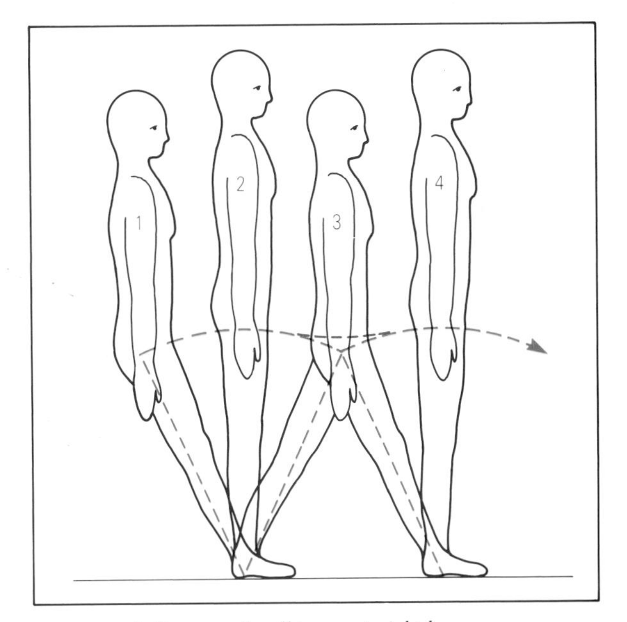

FIG. 5.4 *A diagram of walking on straight legs.*

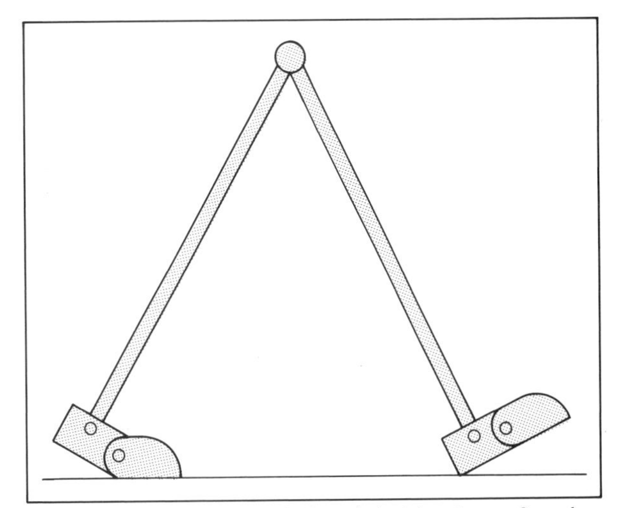

FIG. 5.5 *How we use our heels and the joints in our feet when we walk.*

loses potential energy, but as it falls it moves faster and faster, gaining kinetic energy: most of the potential energy that it loses is converted to kinetic energy (but some is lost as heat, due to air resistance). As a pendulum swings it falls and speeds up, then rises and slows down, as energy is swapped back and forth between the potential and kinetic forms. A frictionless pendulum (if such a thing were possible) would go on swinging for ever, with no need for any fresh input of energy after the first push. Similarly, when you walk, you rise between stages (a) and (b) (FIG. 5.2) but you also slow down because of the way the force on your foot slopes. You fall and speed up between stages (b) and (c). Energy is swapped back and forth between the potential and kinetic forms but the total (kinetic plus potential) energy of the body remains almost constant. A walking person is like a swinging pendulum, so the muscles need do very little work to keep him or her moving.

The objections that have recently been raised to that explanation depend on our improving understanding of muscle physiology. When a muscle shortens while exerting force it does work, which can be converted to potential or kinetic energy. For example, work done by shortening arm muscles can be used to lift a suitcase, giving it potential energy, or work done by leg muscles can be used to give kinetic energy to a sprinter accelerating off the blocks. On the other hand, when a muscle lengthens while exerting force it acts like a brake, converting potential or kinetic energy to heat (for example, when you lower a suitcase slowly to the floor, or decelerate at the end of a race). When you walk at steady speed on level ground your potential and kinetic energies fluctuate during each stride but are the same at corresponding points in successive strides. Any work done by muscles at one stage of the stride must be counteracted by muscles acting as brakes at another.

The problem with explanations of walking that concentrate on the work that has to be done, is this. Experiments on rat muscles like the ones described in Chapter 4 showed that they used oxygen faster when they were shortening, doing work, than when they were holding constant length at the same tension. Other experiments show that muscles use energy less fast when lengthening, acting as brakes, than when holding constant length (again at the same tension). It is not at present clear whether we would use more energy in a style of walking in which our muscles alternately did work and acted as brakes, than one in which they held constant length, doing no work. However, it is quite clear that if the average tension required of the muscles is reduced they will use less energy. That is why I emphasized muscle tension in my account of our straight-legged style of walking.

When a foot leaves the ground at the end of a step its leg must swing forward in readiness for the next step. Metabolic energy will be needed to swing it forward if the job has to be done by muscles, but not if it swings passively like a pendulum.

If you could hold your knee rigid and swing your leg freely from the hip like a pendulum, you would find that, with the knee straight, each swing took about 0.8 seconds (1.6 seconds for a complete cycle). If you allowed it also to bend freely it would swing in a more complicated way, taking 0.7 seconds for the forward swing. It would be very difficult to perform these experiments satisfactorily, but the results have been calculated from measurements on cadavers. They suggest that walking should be particularly easy at speeds that allow 0.7 seconds for the forward swing of each leg.

Much less time than this is available, at normal walking speeds. The forward swing of an adult's leg takes about 0.6 seconds in a slow walk at 0.5 metres per second, falling to 0.35 seconds in a fast walk at 2 metres per second. However, the pendulum calculations assumed that the leg was swinging from a fixed support. Further calculations by Simon Mochon and Tom McMahon of Harvard University showed that the rise and fall of the hip that occurs in walking could make the leg swing forward faster, in 0.35 to 0.6 seconds, depending on how fast the knee is bending at the instant when the foot leaves the ground. This is just the range of swing times that is needed, so the forward swing can be passive. Experiments confirm that there is very little electrical activity in the leg muscles as the leg swings forward.

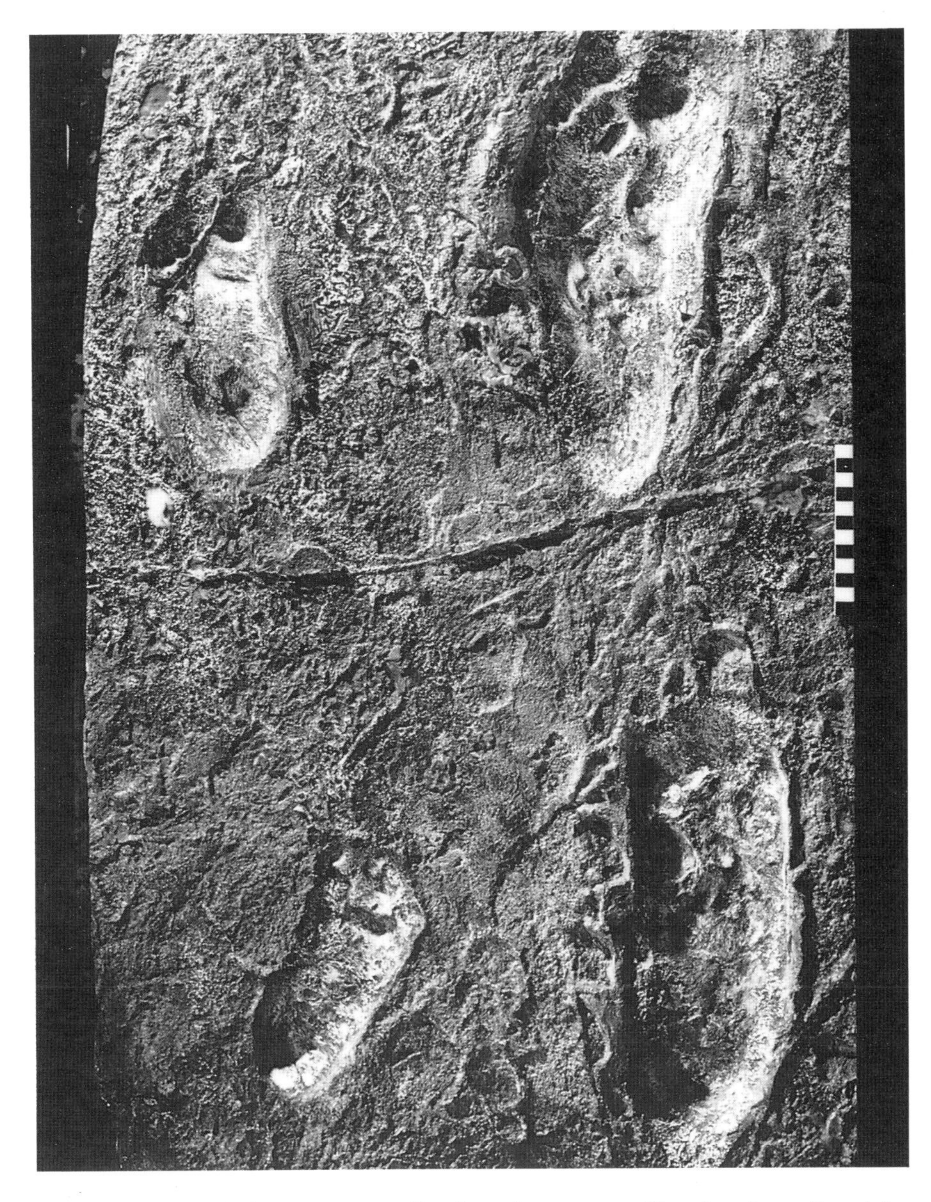

FIG. 5.6 *Fossil footprints at Laetoli, Tanzania. These footprints were made 3.7 million years ago by hominids of about the stature of modern pygmies, who walked essentially as we do. Photo supplied by Dr Michael Day.*

We swing our arms as we walk, for no obvious reason. They swing a little faster than they would swing as pendulums, and electrical recording seems to confirm that they are driven by muscles. The left arm swings with the right leg and vice versa, and it has often been suggested that this helps to keep the trunk facing straight ahead, not turning too much to the left and right in each stride: the arm movements would help to counteract the effects of the leg movements. If this were true our shoulders would turn more from side to side when we walked with our arms by our sides than when we swung them normally, but the reverse effect has been found in experiments.

Our ancestors have been walking as we do for at least 3.7 million years (FIG. 5.6), but apes walk differently. Chimpanzees usually walk on all fours, on the soles of their hind feet and the knuckles of their hands (knuckle walking is a peculiarity of apes), but sometimes walk on hind legs alone, especially when carrying things. Gibbons have immensely long arms that are splendid for swinging through the trees but inconvenient for walking (FIG. 5.7). When they walk on a branch or on the

FIG. 5.7 *A gibbon standing.* © *G. Kinns, Natural Science Photos.*

floor of a zoo cage they use their hind legs only, and hold their hands high.

The pictures of a chimpanzee shown in FIG. 5.8 show how different ape walking is from human walking. These pictures were obtained by X-ray cinematography so they show the movements of the skeleton inside the animal. Notice that the knees bend a great deal, even when the foot is on the ground. Also, the trunk slopes forward. The pictures unfortunately show only a little of the vertebral column but it can be estimated from what is shown that the back was sloping forward at about 45 degrees to the horizontal. This style of walking is very different from the human style, with straight legs and erect back.

Walk a few hundred metres like an ape, with your knees bent and your back sloping. I think that you will find it tiring. The reason is that bigger moments act about the joints than in normal walking, so the muscles have to exert larger than normal forces. In Chapter 4 I explained how straight-legged standing reduces muscle forces and saves energy: straight-legged walking has similar effects.

I have written about apes because they sometimes walk on their hind legs alone, but horses, dogs and other mammals that walk on all fours also walk on more or less bent legs. No animal walks with its legs as straight as we do. That suggests that our walking should be wonderfully economical of energy, in comparison with animals: we will find out whether it is.

We get the energy needed for life and movement from our food. We release the energy when we need it by oxidizing the foodstuffs (in effect, we burn our fuel) and we can find out how much energy a person or animal is using by measuring their oxygen consumption. For every cubic centimetre of oxygen that we use, 20 joules of energy is released from foodstuffs. Conveniently for scientists who want to measure energy consumption, this is approximately true whether the food being used is carbohydrate, fat or protein.

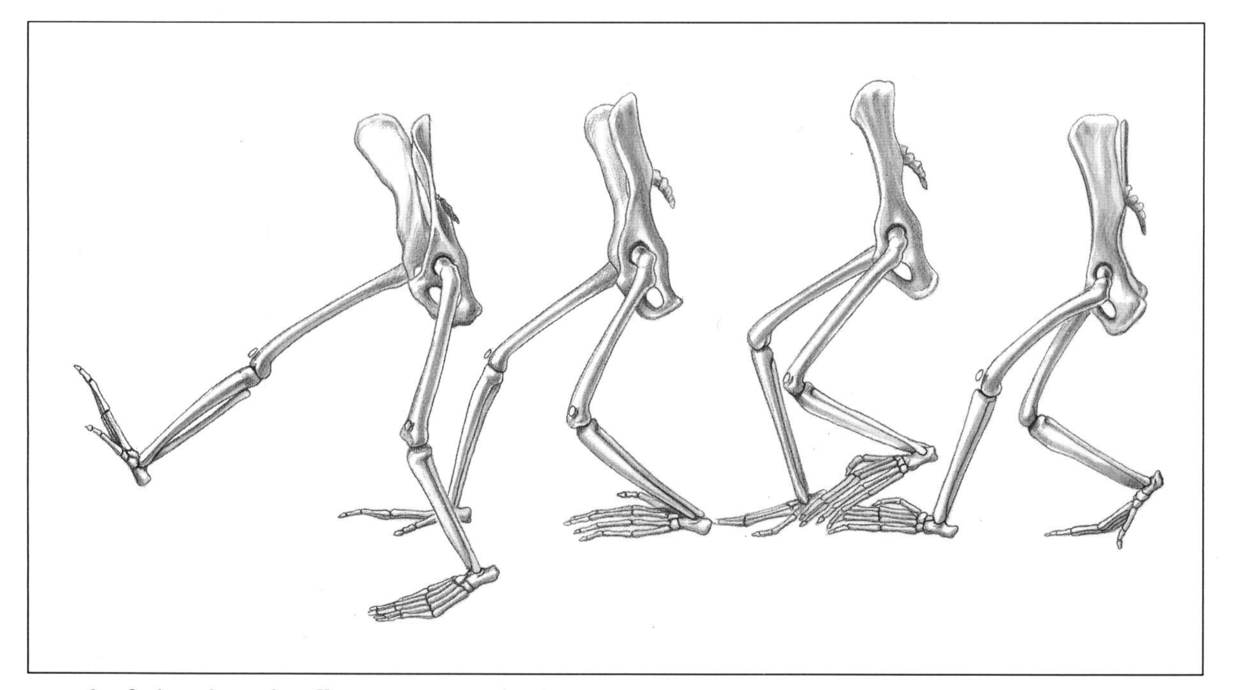

FIG. 5.8 *Outlines drawn from X-ray cine pictures of a chimpanzee walking on its hind legs. From F.A. Jenkins (1972)* Science **178**, *877–879.*

FIG. 5.9 shows measurements of oxygen consumption being made. The woman is walking on the moving belt of a treadmill, walking at the speed of the belt so as to remain stationary relative to the laboratory. One advantage of that is that she can be connected to stationary oxygen analysis equipment. Another is that her speed can be controlled very precisely: if the belt is running at two metres per second, for example, she must walk at exactly that speed to avoid walking off the front of the belt or being thrown off at the back. The air that she breathes out is collected for analysis. Its volume and oxygen content are measured, so it is easy to calculate how much oxygen she has used and from that, how much energy.

Results from such experiments are shown in FIG. 5.10. It shows for example that a man walking at 1.3 metres per second uses energy at a rate of about 300 watts (joules per second). For every metre he travels he uses about 230 joules, or about 140 joules more than if he had remained standing still. Similar measurements have been made on non-human mammals ranging from shrews to cattle, and indicate that a typical human-sized mammal could be expected to use 200 joules per metre more when it walked or ran, than when standing. Human walking at moderate speeds seems to be rather economical, as the straight-legged style led us to expect. However,

FIG. 5.9 *This woman is walking on a moving belt while her rate of oxygen consumption is measured. Photograph by Dr N.C. Heglund, Department of Animal Physiology, University of Nairobi.*

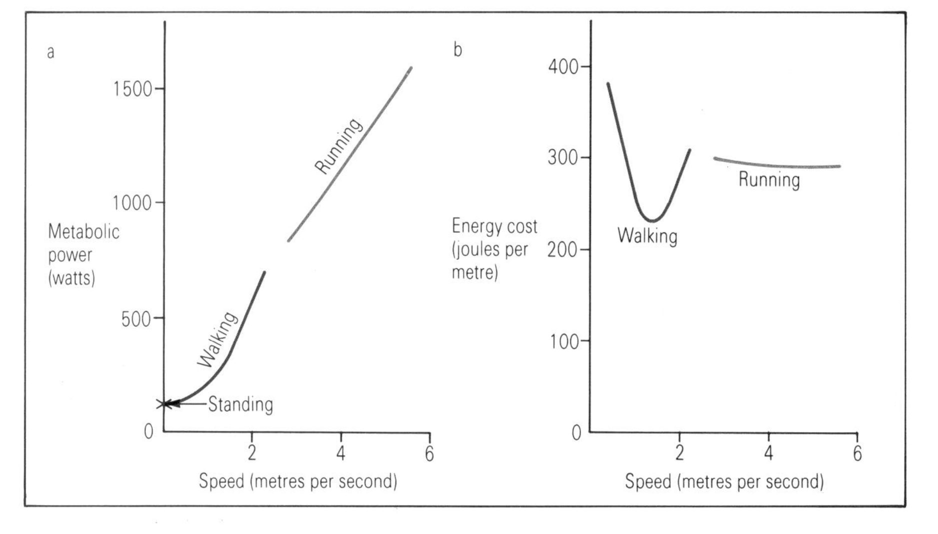

FIG. 5.10(a) *Power consumption (calculated from oxygen consumption) against speed for men walking and running, from the data of R. Margaria (1976)* Biomechanics and Energetics of Muscular Exercise. *Clarendon, Oxford.* (b) *Energy used per metre travelled (same data).*

human running uses about 260 joules per metre more than standing, so our running seems rather uneconomical.

In those experiments, the subjects were free to take long or short strides, as they pleased. In others, people were asked to walk to the beat of a metronome and had to take many short strides per minute, or fewer long strides, depending on the frequency of the beat. It emerged that for each speed there was an optimum stride length, for which the energy consumption was least – and that was the stride length that the people preferred to use, when given the choice. We use short strides to travel slowly and long ones to travel fast, always using the most economical stride length for the speed. The reason that a particular stride length is best at each speed may be that it allows just the right time for each leg to swing forward passively, for the next step.

FIG. 5.10b shows the same data as in (a), plotted in a different way. The energy used per unit time (the power, which is shown in (a)) has been divided by the speed to get energy per unit distance. This graph tells us that if we want to get from one place to another, using as little energy as possible on the journey, we should walk at 1.3 metres per second. Whether or not this is the best speed for walking depends on the circumstances: for example, if you are short of time you will want to walk faster.

Drs Helen and Marc Bornstein are scientists who have attended conferences in many parts of the world. They took advantage of this to observe the speeds at which people prefer to walk, in many different places. Wherever they went, they marked a measured distance on the pavement of the main street and timed people as they walked past. They found that people walk faster in the bustle of big cities than they do in quiet villages, which may not surprise you, but the differences seem remarkably large. In small villages in Greece and Crete, the average speed was about 0.8 metres per second (1.8 miles per hour). In very large cities such as Brooklyn, Prague and Athens it was 1.5 to 1.8 metres per second (3.4 to 4.0 miles per hour).

The energy costs shown in FIG. 5.10 are for travelling over firm, level ground. People use energy up to twice as fast when walking on level, loose sand or swampy ground. (It would be difficult to simulate a swamp on a treadmill. Instead of the apparatus of FIG. 5.9 the people were fitted with large bags, which they carried on their backs while walking over the different surfaces. The air they breathed out was collected in the bags and analysed afterwards.) They also use energy faster when walking up slopes: twice the rate for level ground on a one in 15 slope.

If we want to travel fast we do not walk, but run, and FIG. 5.10b helps us to understand why. It shows energy costs for running, as well as for walking: notice that the graphs for walking and running would cross, if they were extended, at about 2.3 metres per second. Below that speed, walking is more economical than running would be but above it, running is more economical than walking. Given the choice of walking or running, adult people of normal size make the change at about 2.3 metres per second.

Ultrarunning is the sport of racing on foot over a distance of a hundred miles. The best competitors finish the course in about thirteen hours, at a speed of about 3.4 metres per second. They run all the way. Most of the rest take 20 hours or more, at mean speeds of 2.2 metres per second or less, and do not keep constant speed: they alternate between running faster and walking less fast, probably for the following reason. At 2.3 metres per second you can expect to use about 300 joules per metre travelled, whether you walk or run (FIG. 5.10). You can travel the same distance in the same time by alternating between walking at, say, 1.5 metres per second (using 230 joules per metre) and running at 4 metres per second (using 300 joules per metre). The overall energy cost is reduced by alternating between the gaits.

FIG. 5.10 shows that running needs less energy than walking, above the speed of about 2.3 metres per second, but how much faster is it possible to walk, if energy cost is not important? I am going to set out an argument that seems to show that it is not only uneconomic but actually impossible to walk faster than a certain speed.

Rather surprisingly, the argument depends on the dynamics of masses moving in circles. Imagine you tied a brick to one end of a piece of string and then whirled it around your head, holding the other end of the string. You would feel a force in the string pulling on your hand, and if you let go (so that there was no longer a force in the string) the brick would fly off at a tangent. That imaginary experiment tells us that to keep something moving in a circle you need a force (called the centripetal force) pulling it towards the centre of the circle. For an object of mass m moving with speed v around a circle of radius r, the required force is mv^2/r.

The person in the diagram (FIG. 5.11) is walking with speed v on legs of length r. During each of his steps his hip joint moves in an arc of a circle of radius r, centred at the foot. His centre of gravity, a few centimetres above the hip joints, moves in arcs of circles of the same radius. A force mv^2/r is needed, acting towards the centre of the circle, to prevent the person from flying off at a tangent, losing contact with the ground. The feet cannot pull on the ground because they are not glued to it so the only force available to do the job is the person's weight mg. (g is the acceleration of gravity. I explained the meaning of weight in Chapter 3.) If mg were less than mv^2/r the foot would fly off the ground. Thus the style of walking shown in the diagram is possible only if

mv^2/r is less than mg

which implies that v^2 is less than gr

and v is less than $\sqrt{gr}$

A typical adult has legs about 0.9 metres long (from ground to hip joint). The gravitational acceleration is about 10 metres per second per second. Thus $\sqrt{gr}$ is about $\sqrt{(10 \times 0.9)} = 3$ metres per second. This theory says we cannot walk faster than about 3 metres per second, which is only a little above the speed at which running becomes more economical.

If you go out with a small child you will often find that the child has to run to keep up with your walking. The reason is that the child has shorter legs and so has a smaller value of $\sqrt{gr}$. If the legs are 0.5 metres long (a typical length for the legs of a four year old) the limiting speed $\sqrt{gr}$ is only 2.2 metres per second.

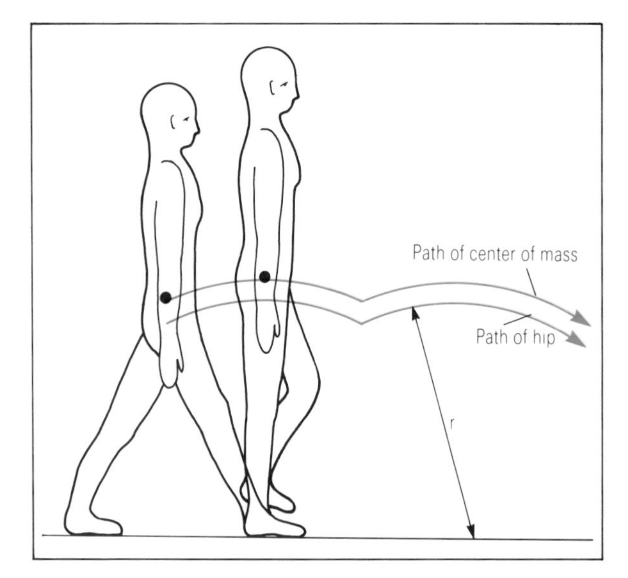

FIG. 5.11 *A diagram to explain the speed limit for normal walking.*

The limiting speed is also low for astronauts on the moon. The gravitational acceleration there is only 1.6 metres per second per second so $\sqrt{gr}$ for an adult is only $\sqrt{(1.6 \times 0.9)} = 1.2$ metres per second. If you insisted on walking on the moon you would have to be content to go slowly. When they visited the moon, the Apollo astronauts actually moved around by hopping like kangaroos (FIG. 5.12).

That argument seems to show that it is physically impossible for normal-sized adults to walk faster than about 3 metres per second, but athletes in walking races go much faster. The men's world record for the ten kilometre walk is 38 minutes 2.6 seconds, giving an average speed of 4.4 metres per second. This is well above our theoretical maximum speed but still much slower than first-rate athletes can run. The men's world record for running ten kilometres is 27 minutes 8.2 seconds, giving a speed of 6.1 metres per second. The women's records are 41 minutes 56.2 seconds for the walk and 30 minutes 13.7 seconds for the run.

The theory seems to say that walking above 3 metres per second is impossible, but athletes manage to walk at over 4 metres per second. Plainly, they cannot do the impossible, but they do not walk in the way that we assumed when we calculated the maximum speed. If you watch racewalkers, or look at photographs like FIG. 5.13, you will see that they wiggle their hips in a very odd way. The pelvis rotates from side to side about a vertical axis, so that the right hip is in front of the left one when the right foot is set down (FIG. 5.14a), but behind the left hip when the right foot is lifted. During the step the supporting hip moves along an arc of a circle from A to B (FIG. 5.14b) but the centre of gravity (between the two hips) travels a little further, from C to D. Also the pelvis rolls about a horizontal axis so that the left hip is higher than the right one while the left foot is on the ground, and vice versa (FIG. 5.14b), so

FIG. 5.12 *Walking would have been very slow on the moon. The astronauts preferred to hop. This is Edwin E. Aldrin. Photo supplied by NASA.*

FIG. 5.13 *Racewalking*. © *Colorsport*.

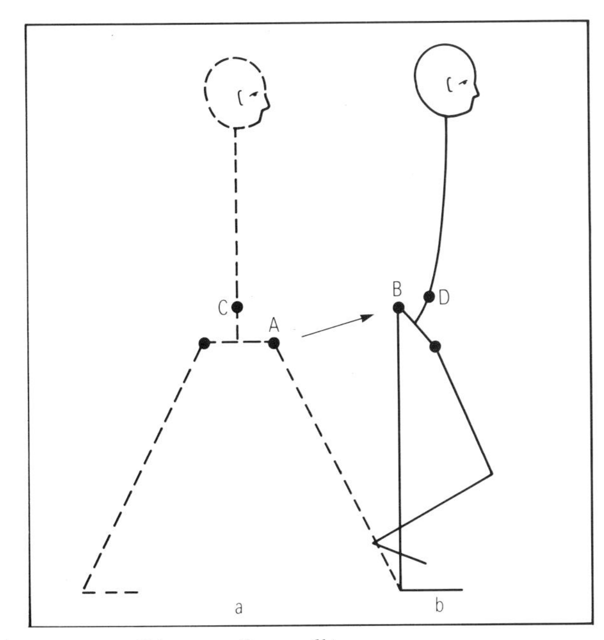

FIG. 5.14 *Diagrams of racewalking.*

the centre of gravity does not rise and fall as much as the hips. The centre of gravity travels further than the hip (while the foot is on the ground) and does not rise and fall so much: in other words, it travels a flatter arc than the hip. The radius of this arc is greater than the length of the leg so $\sqrt{gr}$ is larger and higher speeds are possible.

Let us turn from racewalking to walking on crutches (FIG. 5.15). The body moves forward along arcs of circles much as in FIG. 5.4 but the radius is now the length of the crutches (about 1.35 metres for an adult), not the length of the legs. This suggests that it should be possible to walk faster on crutches than on legs: for a radius of 1.35 metres, $\sqrt{gr}$ is 3.6 metres per second. Actual speeds on crutches are very much slower, but a group of hospital workers who volunteered for an experiment averaged 1.4 metres per second, when travelling on crutches (using only one leg) at their preferred speed out of doors. This is a moderately fast speed, for normal walking.

Paraplegics have paralysed legs because their spinal cords have been damaged. (Our legs are controlled by nerves from the lower part of the spinal cord.) Many of them can be taught to walk on crutches with their legs stiffened by orthoses, but nearly all prefer to use a wheelchair. The reason became apparent when measurements were made of the speeds and rates of oxygen consumption of a group of paraplegics who had been injured at or below the tenth thoracic vertebra. They could propel themselves in wheelchairs at 1.3 metres per second (a moderate walking pace) using oxygen at about the same rate as a healthy person walking at the same speed. Alternatively, they could walk on

FIG. 5.15 *Walking on crutches.*

crutches, but could then manage only 0.5 metres per second and used oxygen a little faster. It was more difficult for them to go fast on crutches than for an amputee who can push off with his one good leg.

Fashions change frequently, but often require women to wear high-heeled shoes. This makes them seem taller, and also affects their gait. Instead of setting their heels down first they set heel and toe down simultaneously (FIG. 5.16b), probably because the foot would be unstable with only a tall, narrow (stiletto) heel on the ground. They take shorter strides, and walk more slowly. For example, a woman in shoes with low, 35 millimetre heels walked at 1.5 metres per second, taking 690 millimetre steps. The same woman in extreme (127 millimetre) stilettos reduced her speed to 0.7 metres per second and took steps only 390 millimetres long. She was accustomed to stilettos but could not stride out in them.

High heels force the foot into a tiptoe position, with the ankle bent toe-down. It cannot be bent beyond a certain angle so the knee has to be bent, to get the toe onto the ground at the same time as the heel. This shortens the stride.

High heels have usually had the narrow stiletto shape but there was a fashion in the early nineteen seventies for very thick high heels (FIG. 5.16a). These shoes often broke at the instep, puzzling manufacturers who had had no trouble with stilettos made of the same materials. It emerged that women walked differently in the thick-heeled shoes. They could land on their heels alone without instability (FIG. 5.16a) and could take long strides. This gait imposed much larger stresses on the shoe than the dainty stiletto gait.

Many walking robots have been built, some of them with practical applications in view. Wheeled vehicles are splendid on smooth roads but get into difficulties on rough ground and are very little use for climbing mountains and staircases. Legged vehicles and animals (for example, mountain goats) can go where no wheeled vehicle could follow. So far,

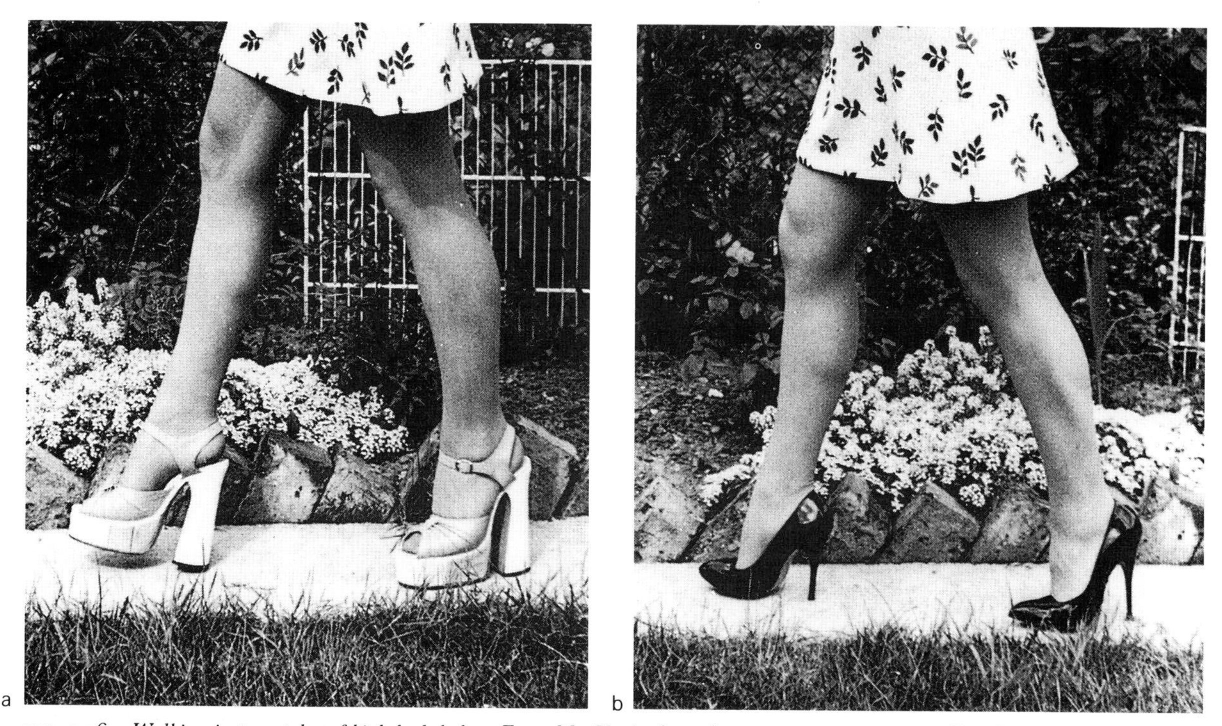

FIG. 5.16 *Walking in two styles of high heeled shoe. From M. Clarke (1975)* J. Brit. Boot Shoe Inst., *Nov–Dec, 171–178.*

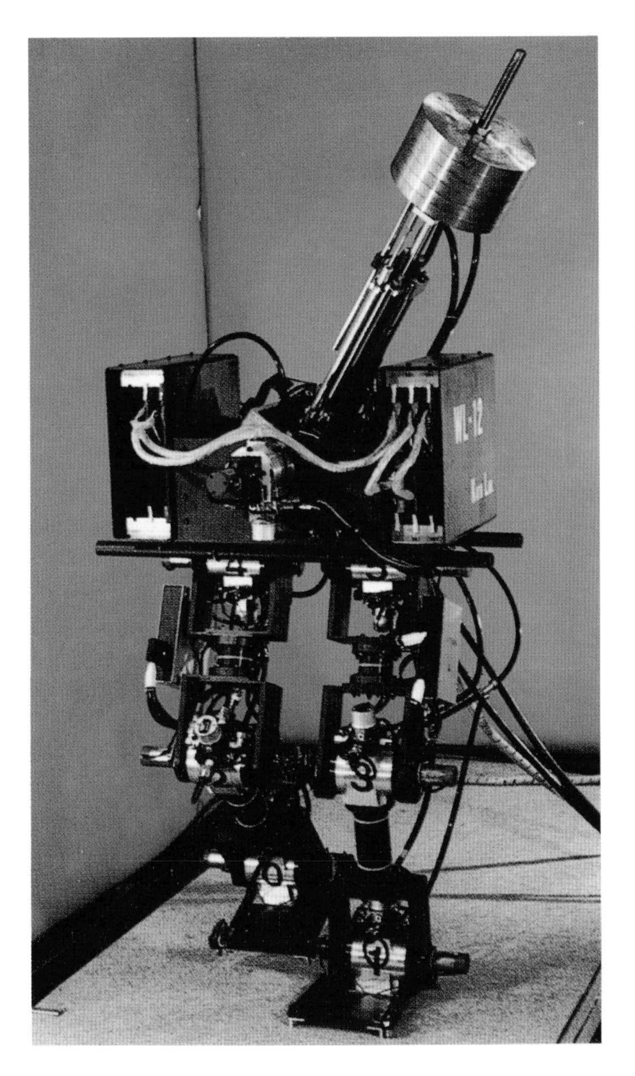

however, very few legged robots have actually been used for practical purposes. One that has is ReCUS, an eight-legged robot that is used for surveying the bottom of the sea.

Many of the other legged robots that have been built walk on two legs like people (FIG. 5.17). One aim is to build robots similar in size and shape to people, that could operate machinery designed for people, in places where people could not be sent. For example, person-sized robots could drive tractors and other machinery into areas contaminated by radioactive materials. We are still a very long way from having robots like that. Most existing two-legged robots are painfully slow and their walk is a clumsy waddle.

So far, economy of energy has not been a major consideration in legged robot design. In any case, the gaits that are economical for robots may turn out to be different from those that are economical for people: a muscle uses energy whenever it exerts a force, even if it is holding constant length, but a mechanical actuator would not necessarily do so. Most two-legged robots walk on bent legs, more like apes than people.

The explanations in this chapter may have made walking seem complicated but I want to end by arguing that the basic principle is really simple, so simple that the old-fashioned toy shown in FIG. 5.18 will walk down a gentle slope. As it goes it rocks a little from side to side, lifting first one foot and then the other. Whenever a foot is lifted it swings forward. You do not have to do anything clever to set this toy going: the appropriate rhythm of movement is established automatically.

FIG. 5.17 *A walking robot, from K. Koganezawa, A. Takanishi & S. Sugano (eds) 1987* Development of Waseda Robot *ed. 2. Kato, Tokyo.*

FIG. 5.18 *A walking toy in the possession of Dr P.J. Evennett.*

CHAPTER SIX

RUNNING

When we walk, each foot is on the ground for more than half the time and, for parts of the stride, both feet are on the ground simultaneously (FIG. 5.2). When we run, however, each foot is on the ground for less than half the time and there are periods when both are off (FIG. 6.1d). Another difference between the gaits is that in walking we keep the knee of the supporting leg almost straight, but in running it bends (FIG. 6.1b). The two gaits are sharply distinct. A walk does not merge into a run, as we increase speed, but changes abruptly within a single stride.

Force plate records show that in running, as in walking, the force on the ground keeps roughly in line with the leg, making the body slow down at stage (a) (FIG. 6.1) and speed up at stage (c). Thus we are travelling slower at stage (b) and faster at stage (d). Also, running is a series of leaps so we are low at stage (b) and high at stage (d). Thus our potential and kinetic energy are *both* high at (d) and low at (b), so the pendulum principle that we saw operating in walking cannot work in running. I will show that much of the kinetic energy that is lost and regained in the course of a step is stored up as elastic strain energy in stretched tendons and ligaments, and returned in an elastic recoil. Running is not like a pendulum, but like a bouncing ball or a child on a pogo-stick (FIG. 6.2).

Most runners hit the ground first with the heel (I will discuss this later) but force plate records show that the force very soon moves forward onto the ball of the foot (FIG. 6.3), under the joints between metatarsals and phalanges (FIG. 4.5). Tension is then needed in the gastrocnemius and soleus muscles to balance the moment of this force about the ankle, and in the quadriceps muscle to balance the moment at the knee. Records from electrodes stuck into the muscles show that they are active for most of the

FIG. 6.1 *Stages of a running stride. Arrows indicate the directions of the forces on the feet.*

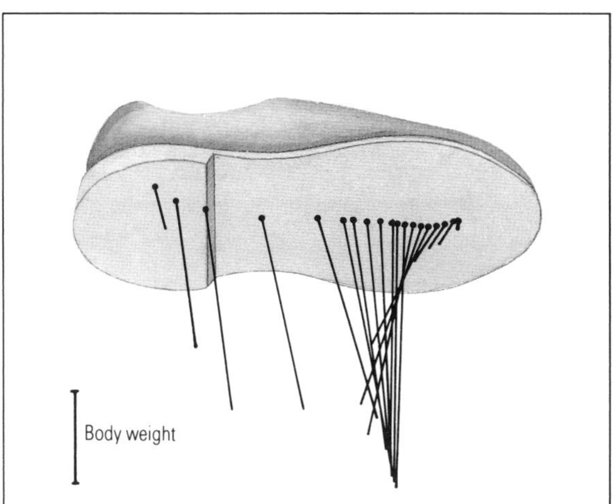

FIG. 6.3 *The forces on a foot during a running step. The lines represent forces at successive intervals of 12 milliseconds, starting at the heel. Each points in the direction of the force and ends at the centre of pressure on the sole. Their lengths are proportional to the forces. After P.R. Cavanagh and M.A. Lafortune (1980)* J. Biomechan. **13**, *397–406.*

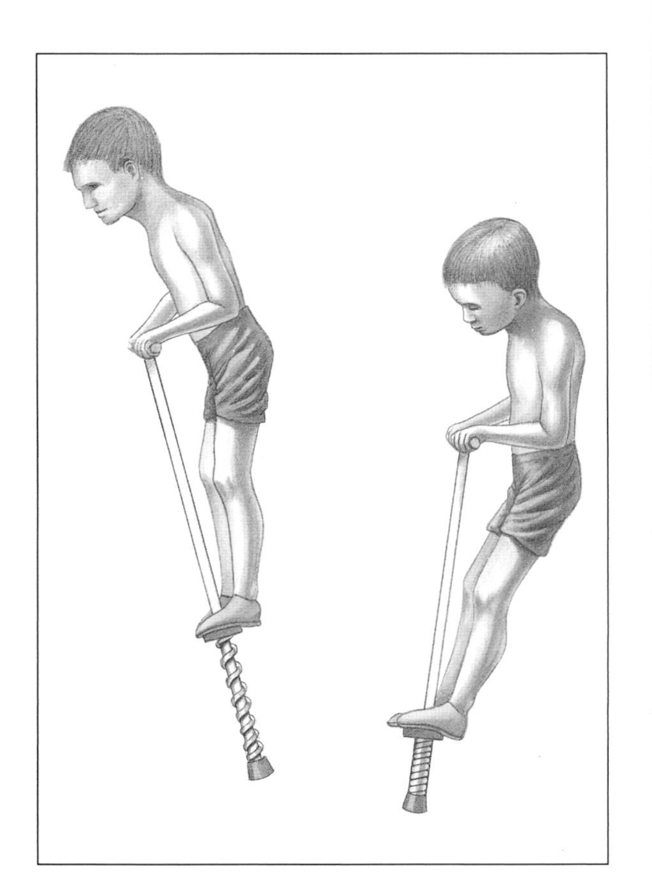

FIG. 6.2 *A pogo-stick spring is compressed on landing and recoils to throw the child back into the air.*

time that the foot is on the ground. They must lengthen as the knee and ankle bend (between footfall and stage (b) in FIG. 6.1) and shorten as the joints extend again (after stage (b)). While they are lengthening, the forces on the ground and in them are increasing, and while they are shortening the forces are decreasing. Springs or pieces of rubber fixed in the same positions in the body would stretch and recoil in much the same way (FIG. 6.4).

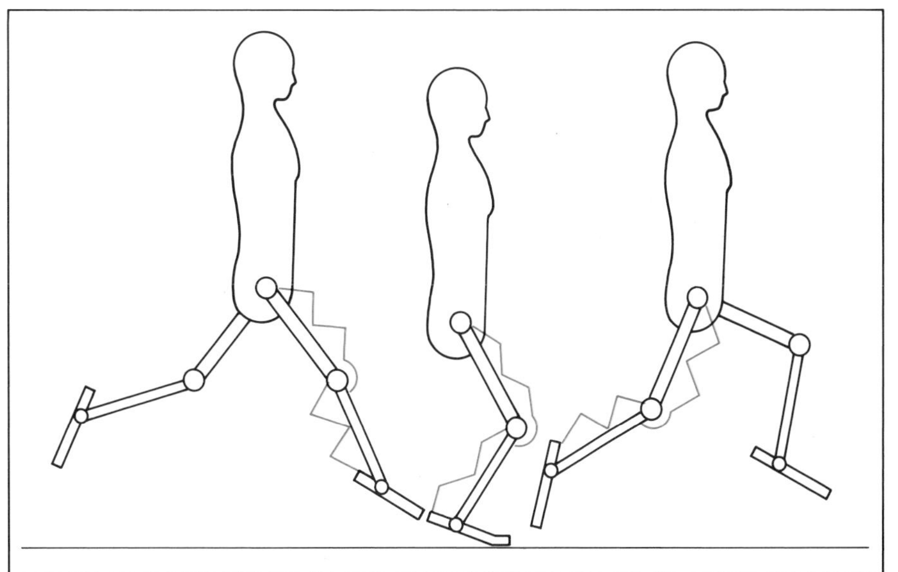

FIG. 6.4 *Major knee and ankle muscles behave like springs when we run.*

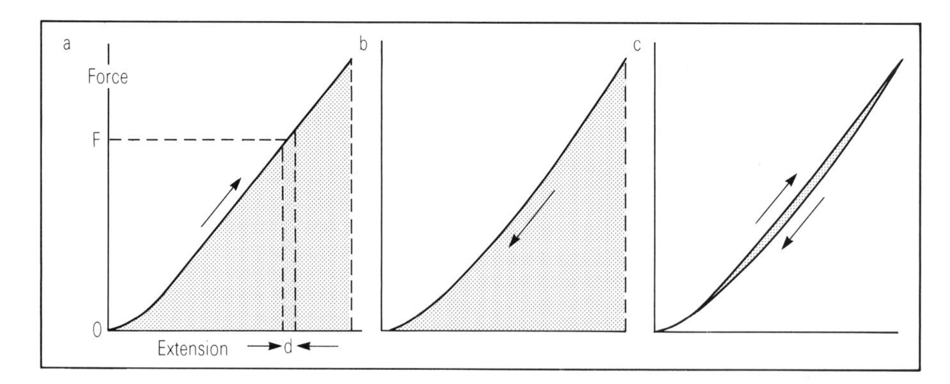

FIG. 6.5 *The record of a tensile test on a tendon, copied from* FIG. 3.5(*b*). *The stippled areas represent in (a) the work done stretching the tendon; in (b) the energy returned by the elastic recoil; and in (c) the energy that was lost.*

Muscle fibres can stretch elastically but only a little, at most by 3% of their length. If they are stretched more, cross-bridges have to detach and reattach, and the elastic strain energy is lost. The most important springs in our legs are not the muscles themselves, but their tendons. The quadriceps muscles have the tendon in which the kneecap is embedded, and gastrocnemius and soleus attach to the heel by the Achilles tendon (FIG. 4.1). Tests like the one illustrated in FIG. 3.5 show that tendons can be stretched up to 8% without breaking, and recoil elastically.

They also show that tendons are very good springs. The record of a test (FIG. 3.5b) is not a single line, but a narrow loop. The force is slightly greater at any particular length while the tendon is being stretched, than when it is shortening. This is because (as with other materials) a little of the energy used in stretching it is lost as heat. We will think about stretching and recoil separately: FIG. 6.5a shows the stretching part of the record only, from FIG. 3.5b. Remember that work is force multiplied by distance (the distance moved along the line of action of the force). That means that if you stretch the tendon by the small amount d shown in FIG. 6.5a the work you have to do is Fd (I have specified a small stretch, so we can ignore the small change in the force F). Fd is also the area of the narrow strip that I have drawn on the graph. You can think of a large stretch as a series of small ones, each requiring work represented by its own narrow strip. That argument tells us that the stippled area in FIG. 6.5a represents the work done in the complete stretch. Similarly the slightly smaller area under the recoil record in FIG. 6.5b represents the energy returned in the recoil, and the area of the loop (FIG. 6.5c) represents the energy lost in the complete cycle. This is only a small fraction of the work of stretching the tendon. Our tests on many tendons show that they return about 93% of the work done stretching them: only 7% is lost as heat. In this respect, tendon is as good an elastic material as most rubbers, though less good than steel springs.

We will calculate how much the Achilles tendon stretches in a running stride. The average force on the ground during the stride must equal the weight of the body, but for much of the time neither foot is on the ground. That means that while the feet are on

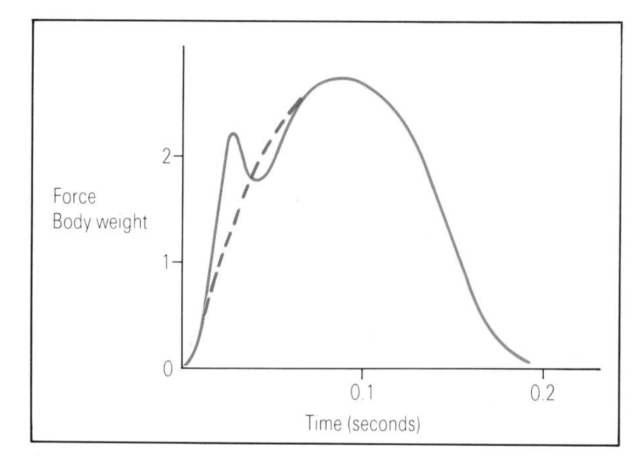

FIG. 6.6 *The vertical component of the force exerted on the ground during a running stride at 4.5 metres per second. This is an average curve calculated from force plate records of five runners. From P.R. Cavanagh and M.A. Lafortune (1980)* J. Biomechan. **13**, *397–406.*

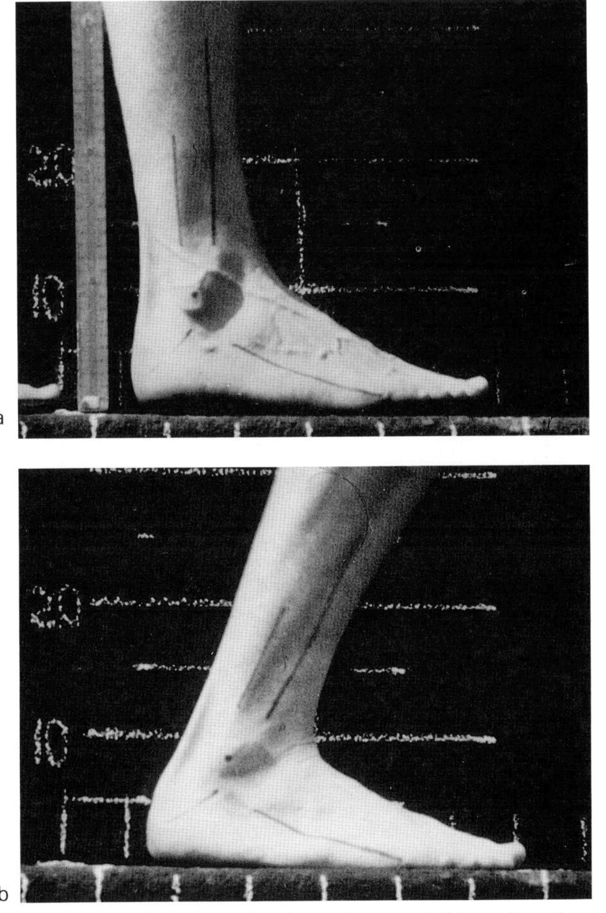

FIG. 6.7 *A foot resting lightly on the ground (a), and at the stage of a running stride at which the peak force acts (b).*

the ground they must exert forces much larger than body weight. FIG. 6.6 shows that they do. It is based on records made at a speed that would give a fairly good time of 2 hours 37 minutes for the 26 miles (42 kilometres) of a marathon race. It shows a peak force of 2.7 times body weight. Larger forces, up to 3.6 times body weight, act in sprinting.

The peak force of 2.7 times body weight (FIG. 6.6) is about 1900 newtons for a 70 kilogramme man. The line of this force is 120 millimetres in front of the ankle joint (FIG 6.8) and the moment arm of Achilles tendon at the ankle is 47 millimetres, so the peak force in the tendon must be 1900 × 120/47 = 4700 newtons, almost half a tonne force. My colleagues and I have measured the cross-sectional areas of Achilles tendons from amputated feet of adult men and find an average value of 89 square millimetres, so the peak stress in the tendon is 4700/89 = 53 newtons per square millimetre. FIG. 3.5 shows that we can expect this stress to stretch the tendon by 6%. It is difficult to be precise about the length of the tendon because it tapers away gradually in the muscle bellies, but most of its fibres are at least 250 millimetres long, so a 6% stretch means 15 millimetres. That is enough to allow the ankle to bend through 18 degrees. A large part of the ankle movement that we see while the runner's foot is on the ground is due to stretching and recoil of the Achilles tendon. The tendon stretches enough for its role as a spring to be very significant.

Notice that the peak stress in this tendon is very much larger than the greatest stress in the thumb flexor, calculated in Chapter 3: it is 53 newtons per square millimetre compared to 12. The percentage length changes are correspondingly larger (6% instead of 2%). The Achilles tendon is much thinner, in proportion to the strength of its muscles, than most other tendons. If it were thicker it would stretch less and be less effective as a spring.

There seems to be another important spring, in addition to the quadriceps and Achilles tendons. FIG. 6.7a shows the foot of my colleague Mike Bennett,

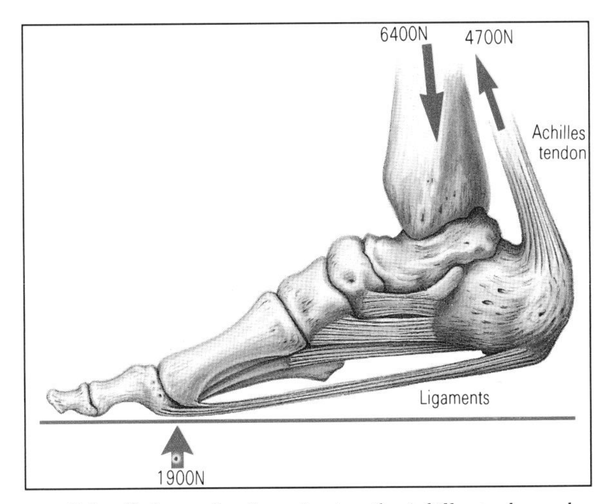

FIG. 6.8 *Skeleton of a foot, showing the Achilles tendon and the ligaments of the arch. The arrows represent the peak forces in a running stride.*

resting lightly on the ground, with most of his weight on the other foot. Notice that the ink mark on his ankle is almost level with the 10 centimetre chalk mark on the wall behind. Photograph (b) shows Mike at the same place at the stage of a running stride at which the ground force is greatest. The mark on his ankle is now almost a centimetre nearer the ground. With other colleagues, he and I were able to show that the foot is a spring which deforms and recoils in each running stride.

The human foot is arched, its arched shape maintained by ligaments (FIG. 6.8). The arrows in the diagram represent forces. I have already explained the peak ground force of 1900 newtons and the Achilles tendon force of 4700 newtons. The third force is the reaction that must act at the ankle joint, to balance the other two. Together, these three forces stretch the ligaments and flatten the arch. That is what has happened to Mike Bennett's foot in FIG. 6.7b.

We showed that the foot really was a spring by experiments in our dynamic testing machine (FIG. 3.4). Obviously, we needed feet that were not attached to people. With permission from their patients, surgeons supplied us with feet that they had had to amputate because of diseased blood vessels. FIG. 6.9a shows how we set up the experiment. The foot is squeezed between the actuator of the machine and the load cell. It rests on metal blocks which in turn rest on a steel plate which is mounted on the actuator. The rollers permit the slight lengthening of the foot that occurs when the arch flattens. The tibia is connected to the load cell. When the actuator moves up the arch is flattened and the load cell registers the force at the ankle joint.

One steel block presses up on the ball of the foot, simulating the ground force. The other presses up on the calcaneus (heel bone): we had dissected the skin and underlying fat off the heel so that the bone rested directly on the steel. This upward push simulated the upward pull of the Achilles tendon. Thus the experiment reproduced fairly well the pattern of force that occurs in running (FIG. 6.8). In most of our experiments we squeezed the foot and let it recoil 2.2 times per second, but changes of frequency made little difference to the results.

FIG. 6.9b shows a typical result. In this particular one the actuator moved up and down eight millimetres (a little less than Mike Bennett's foot was squeezed in FIG. 6.7) and the peak force of 3800 newtons was rather less than the ankle force calculated for running (FIG. 6.8). The arch of the foot flattened and recoiled, acting as a spring, but the area of the loop (FIG. 6.9b) shows that 20% of the energy was lost as heat. The foot is not as good a spring as tendons (7% loss), but it is not too bad.

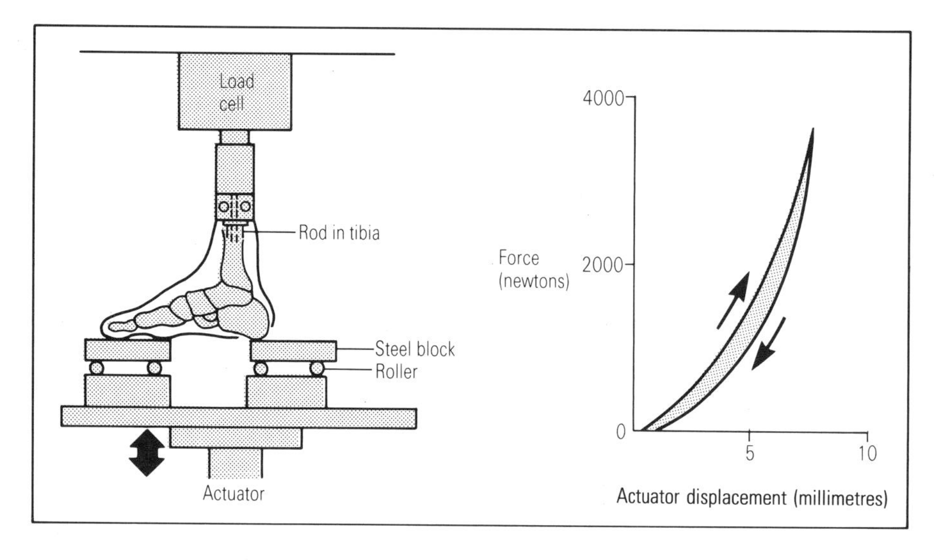

FIG. 6.9 *An experiment on an amputated foot and a record of the result.*

FIG. 6.10 *A running shoe being tested in the dynamic testing machine.*

We will think more about the effectiveness of the springs. We have seen that the body loses and regains potential and kinetic energy while the foot is on the ground. These energies could be measured by analysing film, measuring the changes of height and speed of each part of the body, but it is easier to calculate them indirectly from force plate records. For a 70 kilogramme man running at 4.5 metres per second they total about 100 joules. The elastic strain energy stored in the Achilles tendon and in the arch of the foot can be calculated from the areas under graphs such as FIG. 3.5b and 6.9b. It seems that of the 100 joules, about 35 joules go to stretch the Achilles tendon and 17 joules to flatten the arch of the foot, and are returned (with small losses) in these structures' elastic recoil. Some more must go to stretch the quadriceps tendon, and be returned. Less than half of the 100 joules has to be removed by muscles acting as brakes and returned by them doing work.

The muscles do not necessarily use less metabolic energy because they do less work: they have to exert the same tension, whether the tendons stretch or not. However, the stretching of the tendons and the deformation of the arch of the foot mean that the muscles do not have to lengthen and shorten as much as they otherwise would, and do not have to shorten as fast. It may be that our muscle fascicles are shorter or slower or both, than they would have to be if we did not have springs in our legs. It must cost more metabolic energy to exert tension in a long muscle fascicle, than to exert the same tension in a short one. I explained in Chapter 4 that slow fascicles can exert tension with less cost than fast ones. The springs in our legs enable us to make do with more economical muscles than we would otherwise need.

Most athletes wear shoes which add a little extra spring to their step. Mike Bennett and I have tested running shoes by squeezing the soles in our machine (FIG. 6.10). We tested the part that goes under the ball of the foot, and applied forces of 2000 newtons, imitating the forces on the feet of adult male runners. (Women generally weigh less and exert smaller

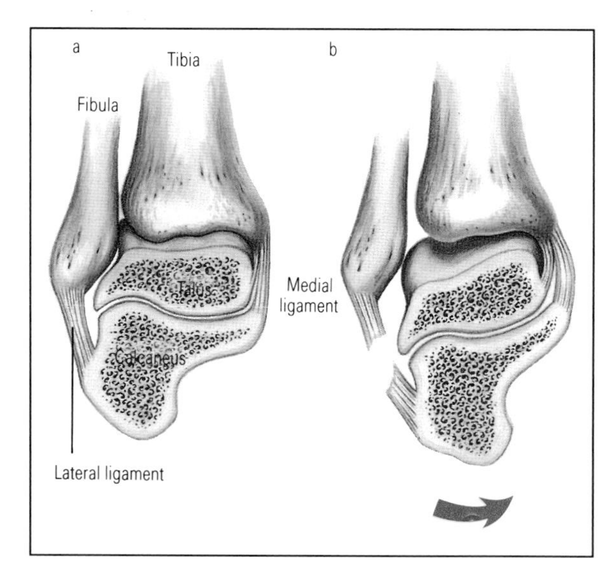

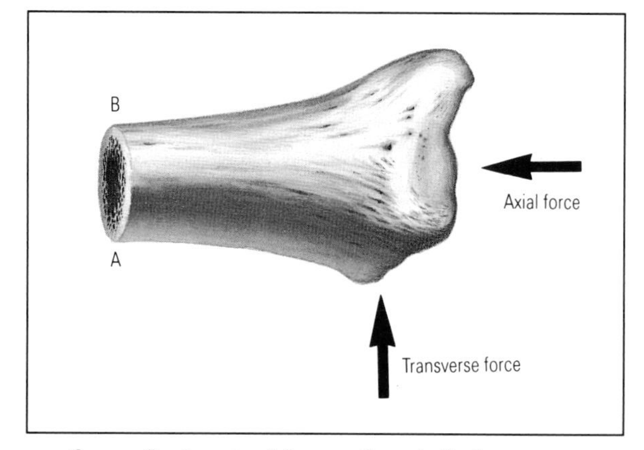

FIG. 6.12 *Components of force on the end of a bone.*

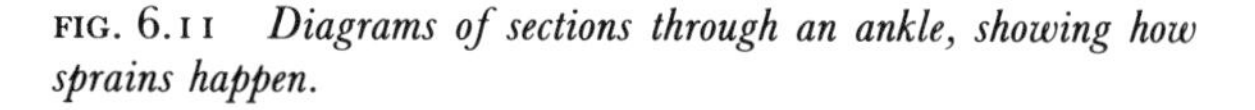

FIG. 6.11 *Diagrams of sections through an ankle, showing how sprains happen.*

forces.) We made the force rise to its peak and fall again in about a quarter of a second, about the time that a runner's foot would be on the ground. We found that most brands were squeezed by about 10 millimetres and stored about 7 joules of strain energy. Thus the sole of the shoe is a less important spring than the arch of the foot which stores (as we have seen) 17 joules. The elastic recoil returned 54 to 66% of the strain energy. Some of the brands were advertised as giving high energy return, but did not necessarily return more of the strain energy than others which were not. In any case, even 66% seems poor in comparison with the arch of the foot which returns 80% of the energy stored in it.

Professor Tom McMahon wondered whether yet more spring might enable people to run faster. If you add a lot of extra elastic compliance by making people run on mattresses, they cannot run fast, but his calculations suggested that a slightly sprung floor might help. With Dr Peter Greene, he designed an indoor running track with its floor mounted on sprung wooden beams. It is much less rigid than conventional tracks: it is pushed down about 9 millimetres, and recoils, at each footfall of an adult runner, storing and returning about 9 joules. It has been a splendid success. Runners regularly run 3% faster on it than on conventional tracks. Remember that 3% is a lot in athletics, representing (for example) 30 metres in a 1000 metre race.

Perhaps it would be possible to get the same advantage on rigid tracks, by using shoes with soles that squeezed by an extra 9 millimetres. These soles would have to be thick but need not be particularly heavy, if the extra compliance were given by bubbles of air. (Some running shoe manufacturers already put gas bubbles in their soles.) Unfortunately running shoe manufacturers tell me that it would be very difficult to make such springy soles satisfactorily because there would be little to prevent the foot rocking from side to side.

Ankle sprains happen when the foot rocks over on its outer edge and large forces act on it: this is particularly likely to happen if you step unexpectedly into a hole. Spraining a joint means tearing a ligament, partially or completely. FIG. 6.11a shows the lateral ligament, which is almost always damaged in ankle sprains, and (b) shows how it gets broken.

Ligaments, tendons and bones need to be strong enough to take the forces that act on them in running and other strenuous activities. We have estimated the peak stress in the Achilles tendon as 53 newtons per square millimetre, in running at 4.5 metres per second. Tests show that the tensile strength of tendon is about 100 newtons per square millimetre, almost twice the working stress. That seems a reasonable, but not over-generous, factor of safety. Engineers commonly design steel structures to be about twice as strong as the greatest loads they are expected to have to bear.

Engineers allow safety factors because you can never be sure that the expected maximum load will not be exceeded, and you cannot be sure that every component will be quite as strong as intended. If structures break frequently, their safety factors are plainly too low. If they never break we should wonder whether they are built extravagantly, with a needlessly large safety factor. People sometimes break their Achilles tendons, usually in the course of sport, but this is relatively uncommon: orthopaedic surgeons with whom I have discussed it estimate that Achilles tendon rupture is of the order of ten times less common than fracture of any particular leg bone. The tendon will heal if the leg is put in a cast but surgical repair is often thought advisable.

The strength of bones raises more complicated problems. Any force on the end of a straight bone can be thought of as a combination of an axial force and a transverse force (FIG. 6.12). The axial force compresses the bone, setting up uniform stresses in every cross section. The transverse force bends it, stretching it at A and compressing it at B, setting up stresses that grade from a maximum tensile stress at A, through zero in the middle of the section, to a maximum compressive stress at B. Transverse forces set up much larger stresses in long, slender structures such as leg bones and sticks, than equal axial forces. For that reason, the easiest way to break a stick is to bend it.

FIG. 6.8 shows the force at the ankle joint at the stage of a running stride when the forces are greatest. The tibia exerts this force on the foot and the foot exerts an equal, opposite force on the tibia. This force is not exactly in line with the tibia, so it has a transverse component as well as an axial one. This transverse component causes the largest stresses in the tibia.

The front surface of the tibia, where the biggest stresses are expected to act, is close under the skin. A member of a research team in Bristol allowed a surgeon to open up his shin and glue a strain gauge to the front surface of his tibia. Strain gauges are small devices that change their electrical resistance when they are stretched, and so can be used to measure the strains (fractional length changes) that occur when structures are loaded. The human guinea pig was sewn up, leaving wires protruding so that the strain in his tibia could be recorded as he walked or ran on a moving belt. The largest value recorded was 0.085%, during (very slow) running. Tests with pieces of bone taken from cadavers had shown that a stress of about 11 newtons per square millimetre was needed to give this much strain, but that the tensile strength of bone is about 150 newtons per square millimetre. That suggests a safety factor of about 14, but that is surely an overestimate because the man ran *very* slowly. Fast running would involve larger forces and a correspondingly lower safety factor. Strain gauge measurements on horses and other animals indicate that their leg bones are generally built with safety factors (for running) between two and five.

However, human runners (and also ballet dancers) often develop small transverse cracks in the front surface of the tibia. They complain of soreness, and the cracks can sometimes be detected in X-ray pictures. They heal if the runner rests but if he or she keeps running they may spread right across the bone so that it breaks. They are sometimes called stress fractures, but that is a bad name because all fractures are caused by stress. They are better called fatigue fractures.

Bone and metals can be broken by repeated application of stresses that would be too small to break them if applied only once. This is the phenomenon of fatigue. For example, in one set of experiments, tensile stresses of 30 to 40 newtons per square millimetre broke bones after an average of about 2000 cycles: the same bones could have stood a single application of 130 newtons per square millimetre. In other slightly different experiments, bones have survived many more repetitions of 30 to 40 newtons per square millimetre, but have again eventually broken. An athlete who runs 100 kilometres each week takes nearly two million running strides each year, stressing each tibia nearly two million times. Healing usually keeps pace with fatigue damage, but if it does not, failure or even complete fracture of the bone may result. Fatigue fractures are common in the tibia, fibula and metatarsals of athletes. They also cause problems in horses, sometimes making them collapse with a broken leg bone in the middle of a race.

FIG. 6.13 *An incident in football that may result in a broken tibia.*

Larger stresses break bones in accidents. In many situations, we are protected from excessive stresses by our muscles. You cannot break my arm by putting gold bricks in the briefcase in FIG. 3.1 because my muscles would be forcibly stretched by loads too small to break the bones. FIG. 6.13 shows how a footballer hit by an opponent may have his leg broken. The victim's foot is braced against the ground, so cannot be protected by muscles giving way.

The bones of old people get broken in much less violent situations, for example when they stumble and fall on a hip. This is partly because bones become more porous and therefore weaker in old age, but mainly because of a change in their composition that makes them more brittle.

Whether things break or not often depends on the energy available to break them. A dropped cup breaks only if its kinetic energy, when it hits the floor, is enough to do the work of breaking it. Breakage of a window depends on the kinetic energy of the stone thrown at it. Brittle materials such as glass need relatively little energy to break them but tough materials such as fibreglass need much more. Fibreglass consists of fine glass fibres embedded in a plastic resin and is tough because a lot of energy gets used deforming and breaking the resin. This is explained in J.E. Gordon's brilliant book *The New Science of Strong Materials or Why you don't fall through the Floor* (Penguin Books, 1968). Bone consists of crystals of calcium phosphate embedded in a mass of collagen fibres, and like fibreglass is tough because of this composite structure. Old people's bones contain a little more than the ideal proportion of calcium phosphate, which makes them brittle. Professor John Currey of York University tested samples of bone from the femurs of people who had died at ages ranging from three to ninety. He found that only one third as much energy was needed to break samples of the oldest bones, as to break equal-sized samples from children. Most of this loss of strength had occurred by the age of sixty.

The long bones of our limbs are hollow tubes (FIG. 3.7b) which gives them strength with lightness. A hollow tube is lighter than a solid rod of the same material if both are the same length and have equal strength in bending. That is why scaffolding is made of hollow tubes. The thinner the wall of the tube, relative to its diameter, the lighter can the tube be made, until its wall becomes so thin that there is danger of it buckling. However, our bones are not empty tubes but are filled with marrow. The thinner the wall of the tube the lighter that is (for given strength) but the greater the mass of marrow that will fit inside it. The total mass of bone and marrow can be made least if the marrow cavity has 0.55 to

0.67 times the diameter of the bone (depending on precisely how you define strength) and many human and other mammalian bones have proportions in that range.

A runner's feet are still moving downwards when they hit the ground, and they are brought to rest suddenly. A big force must be needed to stop them, which explains why force plate records such as FIG. 6.6 show an initial peak of force (in this case, about 25 milliseconds after impact). A quick calculation will serve to check that this interpretation of the initial peak is plausible.

A foot of mass m that has downward velocity v, just before impact, has kinetic energy $\frac{1}{2}mv^2$ that has got to be lost. The foot and shoe are deformed in the impact so the foot is not stopped instantaneously, even on rigid ground, but comes to rest only after travelling a further distance s. As shoe and foot deform, the ground force rises from zero to a maximum F. The work done on the squashed foot and shoe is the mean force (about $\frac{1}{2}F$) multiplied by the distance s: it is $\frac{1}{2}Fs$. This must equal the lost kinetic energy so

$$\frac{1}{2}Fs = \frac{1}{2}mv^2$$

$$F = mv^2/s$$

The mass of an adult human foot is only about one kilogramme but you cannot stop the foot without stopping the lower part of the leg as well, so the effective mass that has to be stopped can be estimated as the foot and lower leg together, about 4.2 kilogrammes. The downward velocity of the foot before impact was found by one investigator to be about 1.5 metres per second, for running at 5 metres per second.

We will make an initial guess that the heel of the shoe and the fatty pad of the runner's heel are together squashed by 12 millimetres, so we will take s to be 0.012 metres. By putting these figures in the equation we can estimate a peak force at impact of $4.2 \times 1.5^2/0.012 = 790$ newtons, about 1.1 times body weight for a 70 kilogramme man. The initial peak in FIG. 6.6 is about 2.2 times body weight but by the time it is reached the rest of the body is being decelerated as well as the foot. The force needed for this has been estimated by drawing a smooth curve rising from zero up to the main peak: adding 1.1 times body weight to it at the appropriate point gives an initial peak like the one in the record. More elaborate calculations confirm the conclusion that the initial peak can be explained as the impulse needed to stop the foot.

Mike Bennett and Robert Ker used our dynamic testing machine to test heel pads from amputated feet and found that forces of 1500 newtons (about as much as typical initial peaks in force plate records of running) squeezed them about four millimetres thinner. Mike and I also tested the heels of running shoes and found that the same force squeezed them by 7 to 15 millimetres. The total deformation of 12 millimetres that we assumed in our calculation was probably not too far wrong.

Impact peaks are suspected of having damaging effects, as causes of osteoarthritis in leg joints. They should be reduced when runners use shoes whose heels have high elastic compliance (which squeeze a lot thinner, under given load) because compliant heels give a large deceleration distance s. Such shoes should give valuable protection, especially on rigid surfaces such as concrete pavements, but it is uncertain whether they do.

Mike Bennett and I found that some running shoes have heels twice as compliant as others, which should make a big difference to impact peaks. Professor Benno Nigg and his colleagues in Calgary made force plate records of athletes running in shoes of different compliances. They found higher impact peaks at higher speeds but at any particular speed there was no significant difference between the impact forces recorded with the various shoes. The runners must have adapted their running style when they changed shoes, but it is not clear how they did this. There was no detectable difference in the velocity at which the foot hit the ground, in the various shoes. There was also no apparent adjustment of knee angle or of the angle of the foot, as it hit the ground. There was some indication that runners in harder shoes hit the ground first with the lateral edge of the heel. This might give a longer deceleration distance because the shoe would have to rotate about its long axis as it settled on the ground.

In another experiment, athletes ran in the same shoes, with and without compliant insoles that were expected to reduce impact forces, but no significant change could be detected.

In contrast to these results, another experimenter found much larger impact peaks when runners were barefoot, than when they wore running shoes. That was when they hit the ground first with the heel, as most runners usually do (FIG. 6.3). If they used a different style, striking with the fore part of the foot, there was no detectable impact peak even when they wore no shoes.

A barefoot runner striking the ground first with his heel might get a nasty jar on rigid surfaces such as asphalt. In a comparison of barefoot running on different surfaces it was found that 54% of first contacts on grass were made with the heel, but only 23% on asphalt. Runners apparently adjust to the hardness of the ground as well as to their shoes.

Athletic injuries are nevertheless more common on less compliant surfaces. 'Jumper's knee' is pain and tenderness below the kneecap, due to rupture of some of the fibres of the quadriceps tendon. It was found in one study that 38% of volleyball players who trained on concrete suffered from it, but only 23% of those that trained on linoleum and 5% of those that trained on wooden floors.

In steady running, forward forces at one stage of the stride balance backward forces at another (FIG. 6.1). To accelerate we must exert mainly backward forces on the ground, so that forward forces act on our feet. A runner who tried to accelerate with his body upright would fall over backwards; he must lean forward to prevent this by keeping the forces on his feet more or less in line with his centre of gravity (FIG. 6.14). The arrows on this diagram represent the principal forces acting on the runner's body, its weight and the force on the foot. (For the present I will ignore air resistance, which is relatively small). These two forces acting together have the same effect as a single resultant force acting at their point of intersection. The size and direction of the resultant can be discovered by drawing a parallelogram of

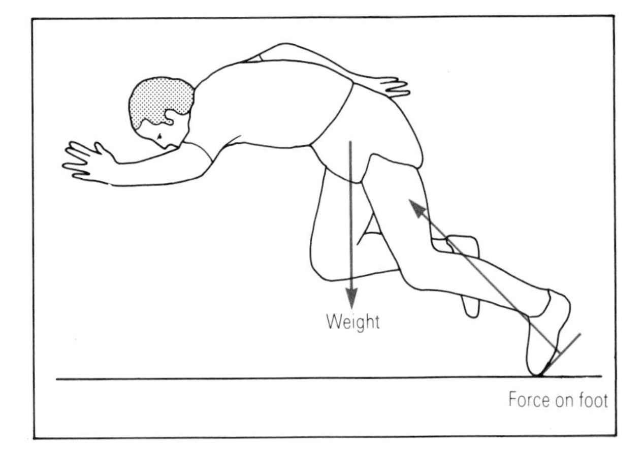

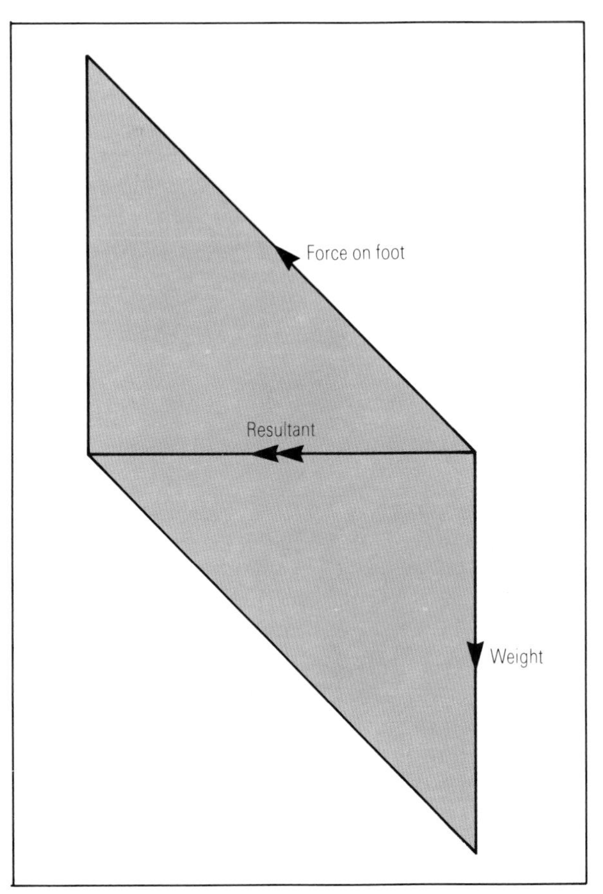

FIG. 6.14 *A sprinter accelerating. The arrows represent forces acting on him.*

forces, as shown in the inset: the sides of the parallelogram are parallel to the forces and their lengths are proportional to their sizes. The average resultant over the first few strides must be horizontal and must act through the centre of gravity, if the sprinter is not to fall over or fly into the air.

The traditional starting position, with the hands resting on the ground, enables sprinters to start the race with their bodies sloping as required for acceleration. Films show that the acceleration over the first few strides may be as much as 10 metres per second per second. For comparison, a Porsche that goes from rest to 60 miles per hour (27 metres per second) in seven seconds has an acceleration of only 3.8 m/s^2. Videotape records made in Tanzania by Dr John Elliott show that lions accelerated towards their prey at about 10 m/s^2 (the same as for human sprinters) but that gazelle and wildebeest accelerated away from them at only 5 m/s^2. This comparison probably reflects the animals' degrees of preparedness, rather than their athletic abilities. The lion and the human sprinter prepare themselves to accelerate but (if the lion creeps up) its prey is taken by surprise. It seems quite possible that if you could put a gazelle under starter's orders and prepare it for the start, it could accelerate much faster.

The horizontal force that you can exert on the ground to accelerate is limited by two things, the strength of the leg muscles and the slipperiness of the ground. The average vertical force on your feet (over a few strides) must equal your weight. The average horizontal force, if you are accelerating at 10 m/s^2, must also equal your weight. (Remember that the downward acceleration of a freely falling body is about 10 m/s^2.) The coefficient of friction between shoe and ground must be at least 1.0, to make this possible. In tests of running shoe friction, five different rubber-soled shoes gave coefficients of about 1.5 on artifical grass, 1.0 to 1.3 on other artificial surfaces and only 0.4 on an artificial surface with loose granules on top. Spiked shoes do not rely on friction but the ground must be firm enough to give them a purchase.

Running round curves must also be limited by the forces required, or by friction, or both. In an experiment, a runner who could manage 10 metres per second in a straight run was unable to go faster than 5.1 metres per second, when running in a circle of 3.7 metres radius. In our discussion of walking I explained that a body moving in a circle needs a centripetal force pulling towards the centre of the circle. If the body has mass m and speed v, and moves in a circle of radius r, the necessary force is mv^2/r. The weight of the body is mg (g is the gravitational acceleration) so the centripetal force is v^2/gr times body weight. This is the mean force towards the centre of the circle, averaged over a complete stride, and the mean vertical force must equal body weight. Thus the coefficient of friction that is needed to prevent the runner skidding sideways is v^2/gr, or 0.7 for running at 5.1 metres per second in a 3.7 metre circle. Runners have difficulty negotiating corners on slippery ground, or on surfaces covered by loose pebbles.

The problem of friction on bends can be eliminated by having a banked track. Look at the arrows representing forces on the feet of a runner negotiating a bend (FIG. 6.15). The single-headed arrows represent the force needed to balance his weight, and the centripetal force. The double-headed arrow represents their resultant, and the runner is leaning to keep the resultant in line with his centre of gravity. Just as an

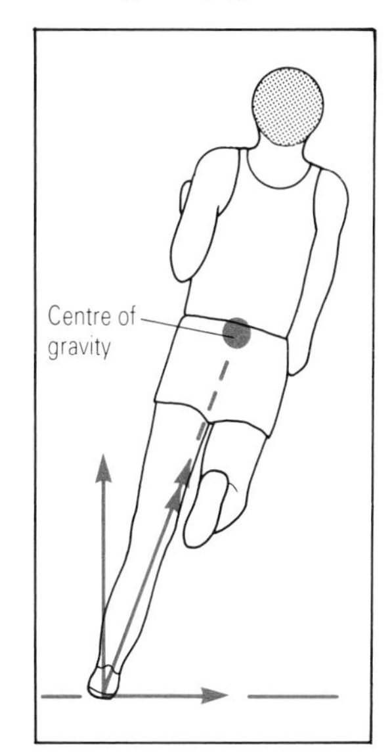

FIG. 6.15 *A runner on a curve. The arrows represent forces acting on the body.*

accelerating runner must lean forward, a runner on a curve must lean towards the inside of the bend. If the track were banked so that its surface was at right angles to the resultant force on the foot, there would be no need for friction to prevent the foot skidding sideways. The angle required would be 36 degrees for the case we have been considering, of running at 5.1 metres per second round a circle of radius 3.7 metres. Shallower banking angles are appropriate for the curves on running tracks, which have much bigger radii.

The resultant force in FIG. 6.15 is greater than body weight. (Remember that the lengths of the arrows are proportional to the forces.) It is only a little greater, but in the extreme case of running at 5.1 metres per second on a 3.7 metre radius it would be 1.23 times body weight. (That is the average resultant force, over a stride.) Peak forces are greater in fast running than in jogging, and speed on curves may be limited by the peak forces needed for running on them.

We will return to acceleration in straight running and consider it from another point of view. The work done by a muscle in a contraction is the force it exerts multiplied by the distance it shortens. The force is limited, and is less if the muscle has to shorten fast (FIG. 3.3b). The amount of shortening is fixed by the movement being made. Therefore the leg muscles can do only a limited amount of work in each step of a sprint start. Film of a (male) sprinter showed that he accelerated from rest to 3.0 metres per second in his first step, to 4.2 m/s in the second step (while the other foot was on the ground) and 5.0 m/s in the third. He gained less speed in each succeeding step, but did his muscles do less work?

He was sprinting on level ground so, ignoring small fluctuations, he was neither rising nor falling and his potential energy was constant. However, he was gaining speed so his kinetic energy was increasing.

$$\text{Kinetic energy} = \tfrac{1}{2}(\text{mass}) \times (\text{speed})^2.$$

Assume that his mass was 70 kilogrammes and calculate his kinetic energy. Initially it was zero. After the first step it was $\frac{1}{2} \times 70 \times 3.0^2 = 315$ joules, after the second 617 joules and after the third 875 joules. In each step he was gaining about 300 joules kinetic energy, which must have come from work done by muscles. The muscles were doing the same amount of work to accelerate the body, in each step. Obviously, this could not go on for ever, or after 100 steps he would be travelling at 29 metres per second (65 miles per hour). This is because the faster a muscle has to shorten the less force it can exert, so the less work it can do. A good sprinter running 100 metres in 10 seconds has an average speed of 10 metres per second but starts from rest, reaches a peak speed of 11 metres per second midway through the race, and slows down again slightly before the finish.

Even this peak speed is unremarkable, by animal standards. Most horse races over distances up to 1600 metres (one mile) are won at 16–17 metres per second and most greyhound races over 460 metres at 15–16 metres per second.

When a runner is travelling at a steady speed the leg muscles do work at one stage of each step and work like brakes at another, as we have already seen. You might suppose that the work and the braking action would balance, but the muscles must do some net work to make good various energy losses. They must do work against friction in the joints, against air resistance and to replace energy lost as heat when tendons are stretched and recoil. The losses in joints are hard to calculate but seem to be small, because the joints are well lubricated. The losses in tendons are also fairly small, only 7% of the energy stored in them in each step. Air resistance is more important when we run fast or into the wind.

Aerodynamic drag (the technical term for air resistance) acts on our bodies both when we move through still air and when air blows past us. It does not matter whether it is the body or the air that is moving: the drag depends on the speed of the one relative to the other. The energy cost of the work done by runners against drag has been measured in ingenious experiments, using a large wind tunnel in an aerodynamics laboratory (FIG. 6.16). The subject runs on a moving belt and their oxygen consumption is measured. A uniform stream of air is driven through the tunnel by a powerful fan. When the fan is turned off, the runner is stationary relative to the air and does not have to do the usual work against drag. When the fan is made to drive the air at the same speed as the moving belt, the runner has to do as much work against drag as when running on fixed ground through still air, at the same speed. As expected, more oxygen was used in the latter case, and even more when the fan was run at higher speeds simulating running into the wind. The experiments were performed only at low running speeds because at high speeds we build up an oxygen debt (as will be explained), making oxygen consumption an unreliable measure of energy cost. However, it was possible to calculate by extrapolating from the results that work done against drag accounts for 7.5% of the metabolic energy cost of running in still air at middle-distance speed (6 metres per second) and 13% at sprinting speed (10 metres per second).

Close behind a runner in a wind tunnel, there is a region of almost still air, shielded by the body. Close behind someone running on fixed ground through still air, there is a region of air moving almost at the speed of the body. In either case, a second runner close behind the first is largely saved the need to do work against aerodynamic drag. Experiments with two runners in the wind tunnel led to the conclusion that a middle distance runner can reduce the metabolic cost of running by 6.5%, by running one metre behind another.

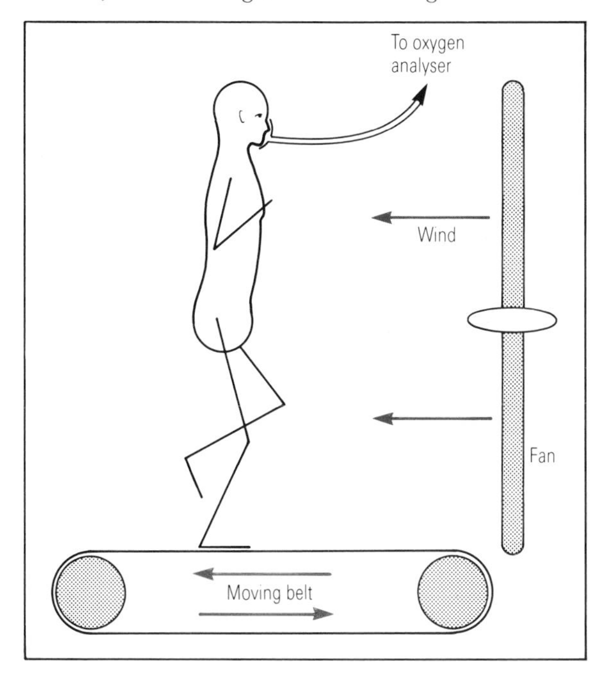

FIG. 6.16 *Running on a moving belt in a wind tunnel.*

The men's world record for running 100 metres corresponds to a mean speed of 10.2 metres per second, and the 200 metre record is almost as fast, but the 1000 metre record is only 7.6 metres per second and the 10000 metre record 6.1 metres per second. The reason is that in long races speed is limited by the rate at which the lungs can take up oxygen and the heart can distribute it round the body. In short races, however, much of the energy comes from processes that do not immediately need oxygen: an oxygen debt builds up, to be repaid later. We cannot continue sprinting for long because only a limited oxygen debt is possible.

A sprinter's first resource is creatine phosphate, a chemical energy source that is kept ready in the muscles for use when the need arises. The concentration of creatine phosphate has been measured in tiny samples of muscle removed by means of a modified hypodermic needle, before and after exercise. Most of it is used in the first 50 metres of a sprint, and replenishment of it is a slow process requiring oxygen.

The second resource is anaerobic glycolysis, a process that releases energy from foodstuffs without any immediate need for oxygen, but produces lactic acid which accumulates in the body. Processes using oxygen can convert the lactic acid back to (a smaller quantity of) foodstuffs, but these processes are relatively slow. Anaerobic glycolysis can supply about half of the energy needed for one minute of violent exercise (a little more than a 400 metre race) but cannot continue indefinitely because the body will tolerate only a limited amount of lactic acid.

When these resources have been used, all our further needs must be satisfied by oxidizing foodstuffs, at a rate that is limited by the properties of our lungs and blood systems. Sprinters need plenty of muscle but long distance runners need big hearts.

CHAPTER SEVEN

JUMPING

Most of this chapter is about the athletic jumping events, the long and high-jumps and the pole-vault, in which the athlete takes off from a run, but I will start by describing standing jumps.

Suppose that you want to jump as high as possible, from a standing start. You bend your legs and then extend them rapidly, to throw yourself into the air. When you stand erect your centre of gravity (50 millimetres above the hip joints) is about 1.0 metres from the ground (FIG. 7.1a). You bend your knees until it is only about 0.65 metres from the ground (FIG. 7.1b) and then extend them rapidly. You rise on your toes at the end of the take-off phase and they do not leave the floor until your centre of gravity is at a height of about 1.05 metres (FIG. 7.1c). Thus you accelerate your centre of gravity over a distance of about 0.4 metres, from 0.65 to 1.05 metres from the ground. Records of athletes jumping as high as possible from force plates show that for much of the acceleration period the force on the ground is about 2.3 times body weight, but it decreases before the feet leave the ground. We will not be far wrong if we estimate the mean force, while the legs extend, as twice body weight. How high can we expect the jump to be?

When a body of weight W is raised by a height h its potential energy is increased by Wh. When a force F moves its point of application a distance s in the direction of the force it does work Fs. The jumper's legs do work which is eventually converted to potential energy as the body rises into the air.

$$Fs = Wh$$

$$h = Fs/W$$

In this case, the force F equals 2 W. The distance s is 0.4 metres. Thus the height gained is $2 \times 0.4 = 0.8$ metres. The initial height (with legs bent) was 0.65 metres so we expect the final height to be about $0.65 + 0.8 = 1.45$ metres, which is about what we observe (FIG. 7.1d).

The peak force may seem surprisingly small. When we run at even moderate speeds we exert 2.7 times body weight with one foot (FIG. 6.6) but in a standing jump *two* feet exert a total of only 2.3 times body weight. Why can we not exert larger forces and jump higher?

Part of the reason is that we bend our knees far more in standing jumps than in running. (Compare FIG. 7.1b with FIG. 6.1b). We have to bend them strongly, to get a large acceleration distance s. In the example shown in FIG. 7.2 the peak force on each foot is only 0.42 times as much in a standing jump, as in running, but the line of action of this force is 1.7 times as far from the knee so its moment (force times distance) is 0.7 times as much. Obviously, that is only part of the story. Now remember how the knee cap mechanism reduces forces and magnifies moments, when the knee is bent (FIG. 4.6). The quadriceps muscles have to exert about 1.3 times as much force to exert the same moment when the knee is bent as in FIG. 7.2b, as when it is straighter as in FIG. 7.2a. The moment in (b) is only 0.7 times the moment in (a) but the force required in the muscles must be about 0.9 times as much as in (a). Finally notice that the ground force exerts a much bigger moment about the hip joint in (b) than in (a), so the hamstring muscles presumably exert bigger forces. They cross the knee as well as the hip, making the moments that the quadriceps have to balance larger than we have estimated so far. Although the force on the foot is so much lower in (b) than in (a), the force in the quadriceps muscles may well be larger.

FIG. 7.1 *A standing jump.*

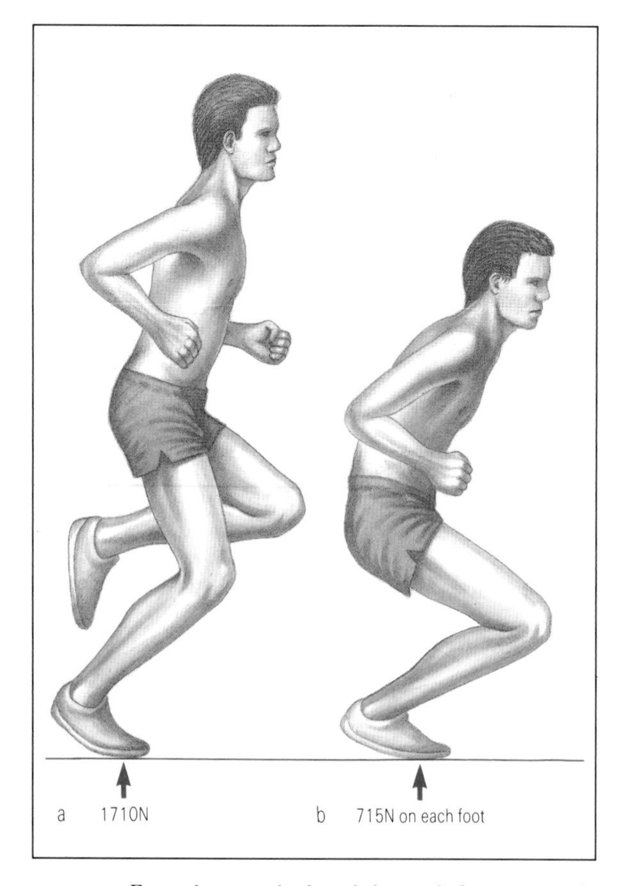

FIG. 7.2 *Force plate records showed that peak force on one of this man's feet was 1710 newtons when he ran but only 715 newtons when he jumped as high as he could. (From the data of R.McN. Alexander & A. Vernon, 1975,* J. Human Movt Stud **1**, *115–123).*

You might think that we should be able to exert larger forces on the ground later in the jump, when our knees are straighter, but by then the muscles are shortening rapidly. The faster they shorten, the less force they can exert (FIG. 3.3).

An athlete who squats for a while in position (b) (FIG. 7.1) cannot jump as high as if he bends and immediately re-extends his legs. In one set of experiments male volleyball players jumped 0.37 metres from a squat and 0.43 metres with the down-and-up action. (In each case the height is the rise after the feet left the floor.) The reason for this has been discovered by experiments on muscles taken from freshly-killed frogs and toads, using apparatus as in FIG. 3.3a. A muscle was held at constant length and stimulated electrically. Then, when it had developed maximum tension, it was allowed to shorten at a controlled rate (FIG. 7.3, line a). The same muscle was stimulated while short, stretched and immediately allowed to shorten at the same rate as before (line b). The force developed in the stretch was greater than when the muscle was held at constant length (FIG. 3.3b) so in the second case the force immediately before shortening was larger. It remained larger than at the corresponding stage of shortening from a stationary start, throughout its contraction. The areas under the curves in this and in other similar experiments show that fast pre-stretching generally increases the work done in a subsequent shortening by about 50%. It is not clear why muscles behave like this, but if human muscle behaves in the same way (and there is evidence that it does) the effect is more than enough to explain why the down-and-up action makes higher jumps possible.

I turn from these small standing jumps to the huge jumps of pole-vaulting (FIG. 7.4). All the work needed for a standing jump is done by the muscles in a single contraction at take-off, but a pole-vaulter

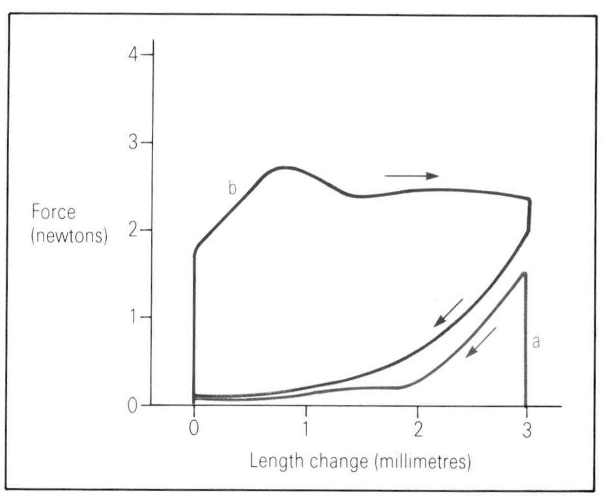

FIG. 7.3 *An experiment with a toad thigh muscle which was allowed to shorten at the same rate from the same starting point, after two different treatments. In (a) it was held for a while at the starting point, while being stimulated, and then allowed to shorten. In (b) it was stretched from a shorter initial length, while being stimulated, and then immediately allowed to shorten. (Based on a record by G.A. Cavagna, B. Dusman & R. Margaria, 1968,* J. appl. Physiol. **24**, *21–32.)*

FIG. 7.4 *Sergey Bubka, the world record holder, taking off for a pole-vault. © Colorsport.*

FIG. 7.5 *Stages of a pole-vault.* © *Colorsport.*

builds up kinetic energy over many strides in a fast run up, and converts that to potential energy for the jump. He or she runs up with a long, flexible fibreglass pole and plants it in a socket in the ground (FIG. 7.5). The pole bends and extends again, lifting the athlete to a great height. The kinetic energy developed in the run up is converted to elastic strain energy in the bent pole which in turn is converted to potential energy at the top of the jump.

Let us see whether that simple theory can explain the heights to which pole-vaulters jump. Consider an athlete of mass m and weight mg (g is the acceleration of gravity) who runs up at speed v and rises in the jump by height h. The kinetic energy at the end of the run up is $\frac{1}{2}mv^2$. At the top of the jump the athlete is moving slowly so has little of this kinetic energy left, but has gained potential energy mgh. If nearly all the kinetic energy has been converted to elastic strain energy which in turn has become potential energy

$$mgh \text{ is almost equal to } \tfrac{1}{2}mv^2$$

sprint speed very important

$$\text{so } h \text{ is almost equal to } v^2/2g$$

Here are some data for a group of good pole vaulters who could jump on average 5.4 metres. (The world record is 6.10 metres.) In their run ups they reached an average final speed of 9.5 metres per second, a little less than the peak speeds of about 11 metres per second reached by fast sprinters in 100 metre races. The gravitational acceleration is 9.8 metres per second per second. Thus the height gained should be almost $9.5^2/(2 \times 9.8) = 4.6$ metres. The athletes' centres of gravity must have been about 0.9 metres from the ground at the end of the run up (slightly lower than in erect standing). They went over the bar with their bodies horizontal, passing their centres of gravity only 0.1 metres over the bar. Thus the centre of gravity should have risen almost to $4.6 + 0.9 = 5.5$ metres and they should have been able to clear almost 5.4 metres. They did indeed clear that height.

The story must actually be a little more complicated. On the one hand, the muscles do some work while the athlete is rising into the air. On the other, some energy must be lost as heat when the pole

4

bends and recoils, and the athlete still has some kinetic energy left at the top of the jump. However the basic story seems fairly accurate. The athlete's kinetic energy is converted first to elastic strain energy and then to potential energy.

Perhaps I should explain how it is possible to get one's centre of gravity as close to the bar as 0.1 metres. FIG. 7.6a shows Alison standing on a narrow piece of wood. She is in balance so her centre of gravity must be vertically above the support: it must be on the broken line. More elaborate experiments show that it is approximately at the point marked by a cross, deep inside the body. If she went over a pole vault bar with the body straight like this, the centre of gravity would be little less than 0.1 metres from the surface of the body, and passing it only that distance above the bar would be tricky. However, pole-vaulters actually go over the bar with their arms raised and the body bent forward at the waist, rather as in FIG. 7.6b. In this position, as in the other, the centre of gravity must be vertically above the support. It is slightly in front of the skin of the belly, making it much easier to pass it close over a bar. In case you find it difficult to see how the centre of gravity can be outside the body, think of a wedding ring. The centre of gravity must be in the empty space in the centre of the ring, not in the gold itself.

An important part of the skill of pole-vaulting is adjusting the position of the body so as to clear a bar very close below the centre of gravity. In the experiment that I described, the pole-vaulters who could clear 5.4 metres passed their centres of gravity only 0.10 metres above the bar, but a less able group who could only clear 4.9 metres passed theirs 0.23 metres above the bar. (The less able jumpers also ran up less fast than the better ones).

Now we will think about ordinary high-jumping, without a pole. At the end of the run up, the athlete sets a foot down well in front of the body, with the knee almost straight (FIG. 7.7). The knee bends and re-extends and the athlete is thrown into the air. The

FIG. 7.6 *Balancing on a narrow bar. The dots show the approximate position of the body's centre of gravity.*

FIG. 7.7 *Taking off for a high-jump. Illustration supplied by Dr J. Dapena.*

leg behaves rather like a pole-vaulter's pole.

Imagine that the leg were a spring that took up all the jumper's kinetic energy, stored it as elastic strain energy and returned it in an elastic recoil. Imagine also that the jumper ran up as fast as a pole-vaulter. The forces on the take-off leg would be enormous, far more than on a pole-vaulter's pole, because the jumper would be decelerated and re-accelerated in a very short distance. However, if the legs were strong enough to stand these forces the jumper would be thrown into the air with enough velocity to rise as high as a pole-vaulter.

C

High-jumpers actually jump far less high: the men's world records are 6.10 metres for the pole-vault and only 2.44 metres for the high-jump. They run up much more slowly than pole-vaulters (I will try to explain why) and their legs do not work as perfect springs. A good high-jumper's speed at the end of the run up is about 7.0 metres per second, so he has less kinetic energy than a pole-vaulter who runs up at 9.5 metres per second, but even this amount of energy would be enough to raise his centre of gravity by 2.5 metres, from its initial 0.9 metres to 3.4 metres, if it could all be converted to potential energy. That would enable him to clear 3.3 metres, far above the world record.

Part of the discrepancy is due to the athlete's not losing all his kinetic energy: he still has a forward speed of about 4 metres per second as he goes over the bar. Here are some data for men filmed during a competition in which they cleared an average of 2.04 metres. Just before take-off they were running at a mean speed of 6.7 metres per second. Their mean mass was 76 kilogrammes giving a kinetic energy of $\frac{1}{2} \times 76 \times 6.7^2 = 1706$ joules. As they crossed the bar their mean speed was 4.2 metres per second, giving a kinetic energy of $\frac{1}{2} \times 76 \times 4.2^2 = 670$ joules. Their centres of gravity rose from a height (at the end of the run up) of 0.91 metres, to about 2.10 metres as they crossed the bar, so the gain of potential energy was $76 \times 9.8\ (2.10 - 0.91) = 886$ joules. That leaves $1706 - 670 - 886 = 150$ joules unaccounted for. Some of this, probably about 50 joules, can be explained by the body rotating as it

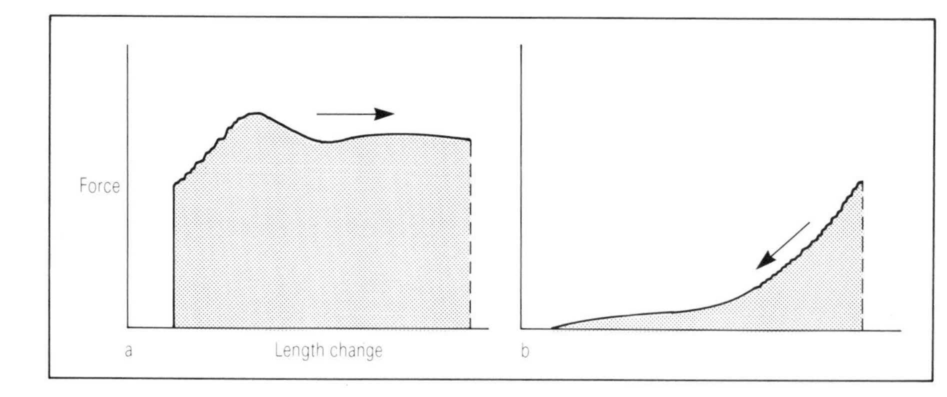

FIG. 7.8 *Tracings from record (b) of* FIG. 7.3. *The stippled areas represent energy dissipated as the muscle was stretched and work done by it as it shortened.*

OPPOSITE PAGE

FIG. 7.9 *The straddle.* © *Colorsport.*

FIG. 7.10 *The Fosbury flop.* © *Colorsport.*

flies through the air. (Rotating bodies have kinetic energy even if their centres of gravity are not moving: think of a flywheel.) However, about 100 joules of the athletes' initial kinetic energy must have been dissipated as heat.

That must have happened in muscles, which can exert much larger forces when being forcibly stretched, than when shortening (FIG. 3.3). Record (b) in FIG. 7.3 shows the force exerted by a fully activated muscle which was stretched and then allowed to shorten at the same speed. It exerted a braking effect as it was stretched, dissipating energy proportional to the stippled area in FIG. 7.8a. As it shortened again, it did work proportional to the stippled area in FIG. 7.8b. The energy dissipated as heat when it was stretched was far more than the work it did as it shortened. In steady running the braking and accelerating effect of the muscles are in balance (the muscles are not fully activated during the braking phase), but in take-off for a jump the muscles have to be fully activated throughout to supply the large forces required, so the braking forces are much larger than the accelerating ones.

You might think that high-jumpers could jump higher if they ran up faster – but if they could they presumably would. Also, a mathematical model that takes account of the physiological properties of muscle seems to show that a faster run up would actually give a less high jump. The reason seems to be that the force muscles can exert is limited: no matter how fast a leg muscle is being stretched it cannot exert more stress than about 0.5 newtons per square millimetre (FIG. 3.3). The mathematical argument is hard to put into words, but here is the gist of it.

The body travels horizontally in the run up, but to rise as high as possible it must be thrown into the air with the greatest possible upward component of velocity. Its upward acceleration is proportional to the unbalanced upward force on it: that is, to the upward force on the foot minus body weight. The upward velocity that it gains is the upward acceleration multiplied by the time during which it is being accelerated. If the force is limited, the upward velocity at take-off depends mainly on the acceleration time. The foot cannot stay at one point on the ground for long if the body is travelling fast, so if it goes too fast the acceleration time is short and the upward velocity low.

High-jumping, like pole-vaulting, is not merely a matter of raising the centre of gravity as high as possible. A large part of the skill is adjusting posture so as to clear the highest possible bar, for a given rise of the centre of gravity. In the straddle technique (FIG. 7.9) jumpers go over the bar face down, and position their legs so as to bring the centre of gravity out in front of the belly (as in FIG. 7.6b). In the Fosbury flop technique, introduced by Dick Fosbury in 1968 (FIG. 7.10), the jumper faces upwards and bends backwards, to bring the centre of gravity out behind the back (rather as in FIG. 7.6c). In the straddle the centre of gravity passes well over the bar, usually about 150 millimetres over it, but flop jumpers occasionally succeed in passing it just under the bar.

cantabrian
NIKE

Long-jumpers, like high-jumpers, set down a foot at the end of the run up, with the knee almost straight (FIG. 7.11). The knee bends and extends, and the jumper takes off. There are two striking differences from the high jump take off. First, long-jumpers run up extremely fast, at speeds up to 10.5 metres per second: this is very close to the peak speeds of outstanding sprinters. Secondly, they do not set the foot down as far in front of the body. The angle of the leg to the horizontal, when the foot is set down, is 60–65 degrees for long-jumpers but 45–50 degrees for high-jumpers.

The reason is this. High-jumpers are concerned only to gain as much height as possible. Long-jumpers also want to gain height but want to keep as much of their forward speed as possible, to travel as far as possible before they fall to the ground. If they set down the leg at too shallow an angle, the ground force (acting more or less in line with the leg) slows them down too much. They do best to run up fast and set the leg down at a steep angle.

Huge forces act in long-jumping, just after the foot strikes the take-off board. Force plate records of Finnish national long-jumpers showed peaks of 12 times body weight (at 68 degrees to the horizontal). This is far above the peaks of about three times body weight (at 90 degrees) that are usual in running (FIG. 6.6).

Ancient Greek long-jumpers used a technique that would be illegal now. They carried weights called halteres which they held in front of the body as they took off but behind as they landed (FIG. 7.12). While the jumper was in the air he could do nothing to alter the path of his centre of gravity. (I am ignoring the very tiny changes of speed that he could make by altering his position so as to change aerodynamic drag.) However, he could change the position of his centre of gravity relative to his feet. Suppose that an athlete carried a pair of halteres which together have one ninth the mass of his body. (This is within the range suggested by museum specimens which have masses of two to nine kilogrammes the pair.) The centre of gravity of athlete plus halteres would then be one tenth of the way along a line joining the centre of gravity of the body to that of the halteres. If the athlete moved the halteres one metre backwards,

FIG. 7.12 *An ancient Greek vase showing a long-jumper landing. All rights reserved, The Metropolitan Museum of Art.*

FIG. 7.11 *Long-jumping. After G. Dyson (1973)* The Mechanics of Athletics *ed 6. University of London Press, London.*

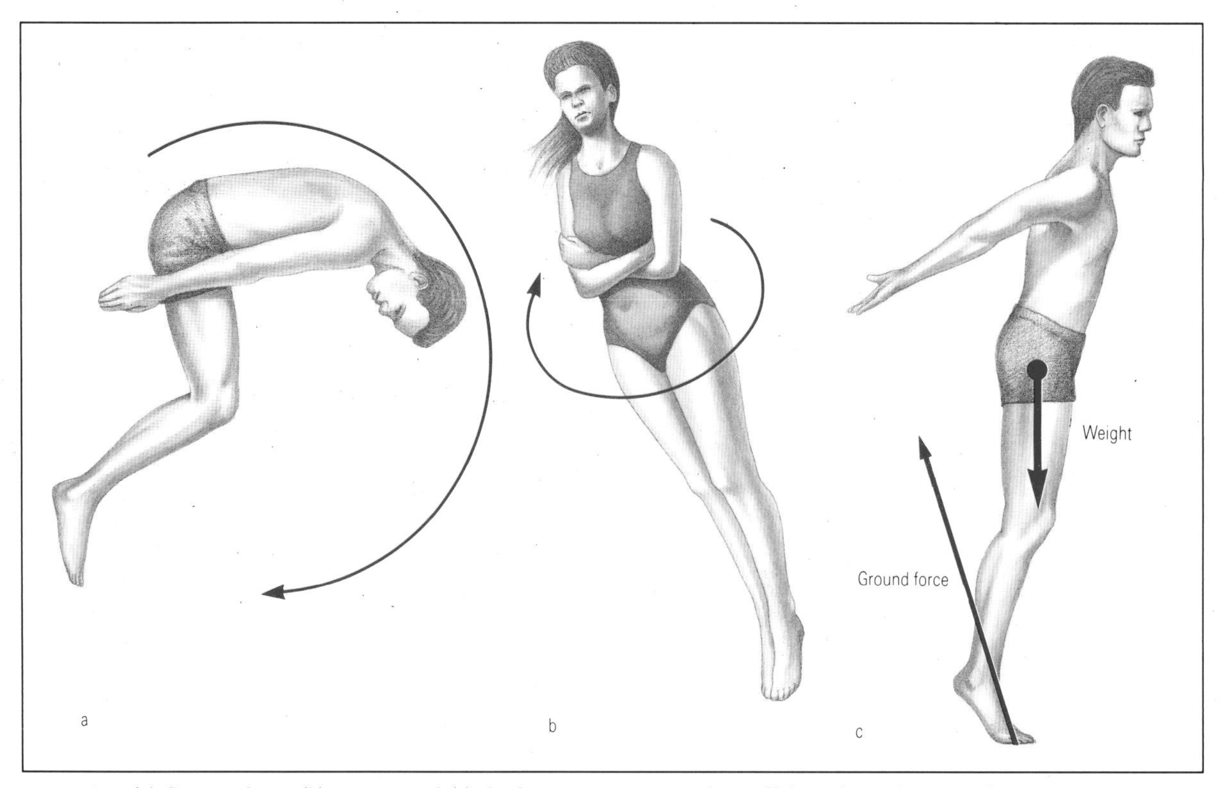

FIG. 7.13 *(a) Somersaulting, (b) twisting and (c) the forces on a gymnast taking off for a forward somersault.*

relative to his body, he moved them 0.9 metre backward and his body 0.1 metre forward, relative to the combined centre of mass. That way he could add 0.1 metre to his jump.

He could add more if he threw the halteres away behind him, while in the air. Remember Newton's second law of motion, force equals mass multiplied by acceleration. If the jumper exerted a backwards force on the halteres to accelerate them backwards, they would exert an equal forward force on him accelerating him forwards. (Similarly, rockets work by accelerating gases backwards.) If the forces are equal, the accelerations must be inversely proportional to the masses: the forward acceleration of the 70 kilogramme athlete would be one ninth of the backward acceleration of the 8 kilogramme halteres. The forward velocity of the athlete would be increased by one ninth of the change of velocity of the halteres, and he would jump correspondingly further. Ancient Greek athletes seem not to have used this technique: it was presumably against the rules.

This is an example of the law of conservation of momentum. The momentum of a body is its mass multiplied by its velocity. If no external force acts on a group of bodies, their total momentum remains constant. The law can be applied approximately to the horizontal component of the velocity of jumper and halteres because the only horizontal force on them, air resistance, is too small to change their momentum much during the jump.

We will leave field athletics now, and look at the complicated rotations that trampolinists and divers make, while flying through the air. Somersaulting means rotating about an axis that runs transversely to the body (FIG. 7.13a). Twisting means rotating above the long axis of the body (FIG. 7.13b). FIG. 7.13c shows a gymnast taking off for a somersault. The force on his feet is not in line with his centre of gravity so it will set his body rotating as it throws him into the air. Similarly, he could set himself twisting as he jumped by exerting a forward force with one foot and a backward force with the other. Trampolinists and divers often make twisting somersaults (FIG. 7.14). The jump or dive may start as apparently simple somersaulting (without twist) and the twisting may start suddenly in mid air. How

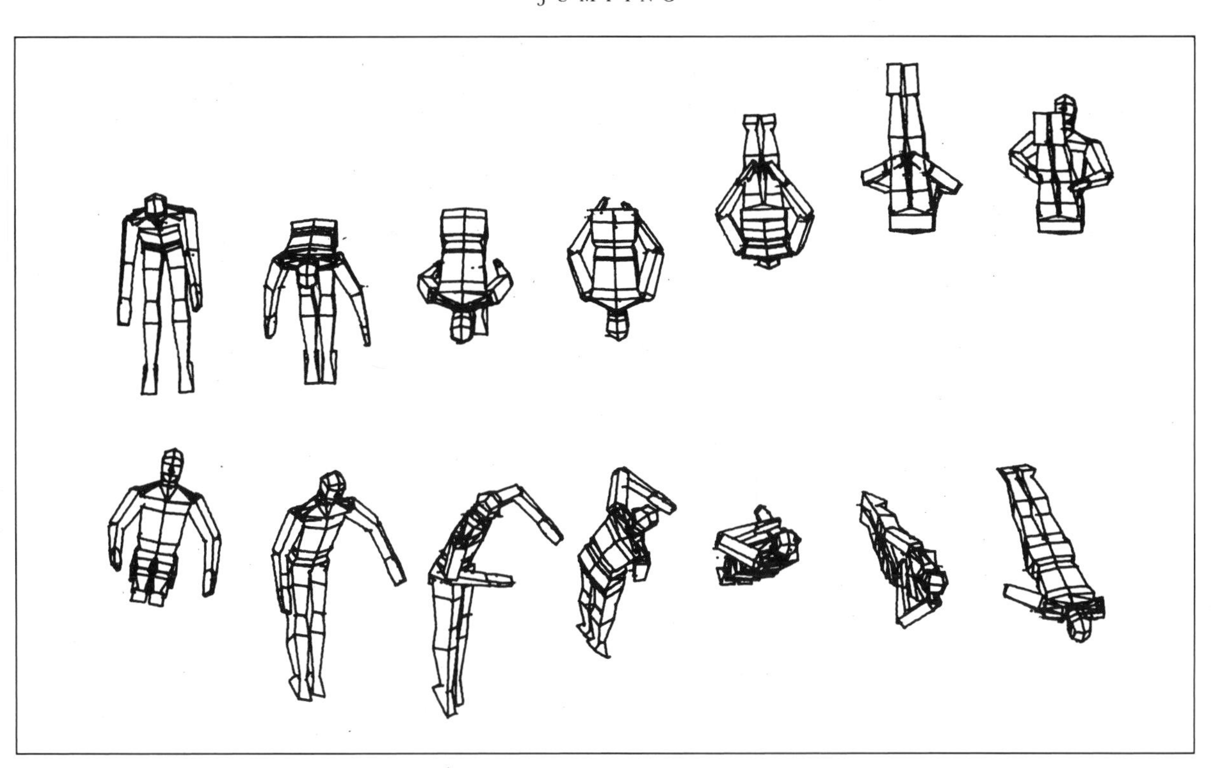

FIG. 7.14 *A double forward somersault with $1\frac{1}{2}$ twists in the second somersault: a computer graphics sequence derived from a film. From M.R. Yeadon (1984)* The Mechanics of Twisting Somersaults *Ph.D. thesis, Loughborough University.*

is it done?

The explanation involves the concept of angular momentum. Remember that momentum (the ordinary linear sort that came into the haltere discussion) is velocity multiplied by mass. Angular momentum is angular velocity (revolutions per second, if you like, but scientists use the units called radians per second) multiplied by moment of inertia, which depends on the masses of all the parts of the body and how far they are from the axis of rotation. The moment of inertia of the human body for somersaulting is much larger than the moment of inertia for twisting, because much of the body is a long way from the somersaulting axis but close to the twisting axis. The law of conservation of angular momentum says that the angular momentum of a group of bodies remains constant, unless external moments act on it. No moments (apart from small ones due to air resistance) act on divers' bodies in mid-air so no matter how parts of the body move relative to each other the angular momentum of the body as a whole remains constant. That is the clue to understanding twisting somersaults.

Imagine someone flying through the air, neither somersaulting nor twisting (FIG. 7.15a). He has no angular momentum, and he cannot give himself any. If he rotates his arm anticlockwise (as seen in the diagram) his body must rotate clockwise so as to have equal and opposite angular momentum: raising the arm through 180 degrees (as shown) makes the body tilt about 9 degrees. Now suppose that the person makes the same arm movement while somersaulting. Computer simulation shows that this starts him twisting (FIG. 7.15b). I will try to explain why.

Initially the person is somersaulting, rotating about the axis shown in FIG. 7.16a. Her angular momentum can be represented by an arrow along the axis of rotation. (By convention, the arrow points in the direction that makes the rotation clockwise, as seen when looking in the direction that the arrow points.) Raising the arm tilts her body but the angular momentum remains unchanged, so the arrow representing it is no longer at right angles to the long axis of the body (FIG. 7.16b). Angular momentum can be broken down into components in different directions, like forces and velocities. In

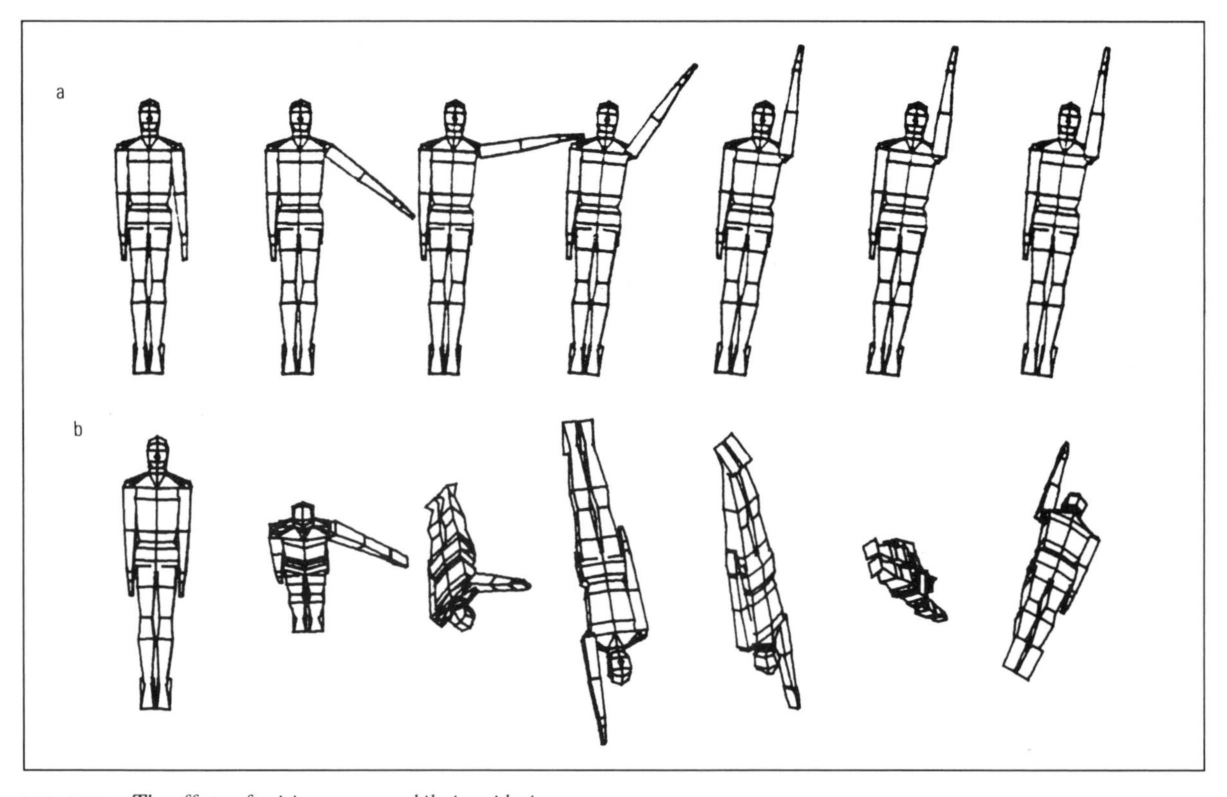

FIG. 7.15 *The effects of raising an arm while in mid air, (a) without and (b) with somersaulting. These diagrams were generated by a computer programme which calculated the movements, taking account of the masses and moments of inertia of the segments of the body. From M.R. Yeadon (1984)* The Mechanics of Twisting Somersaults *Ph.D. thesis, Loughborough University.*

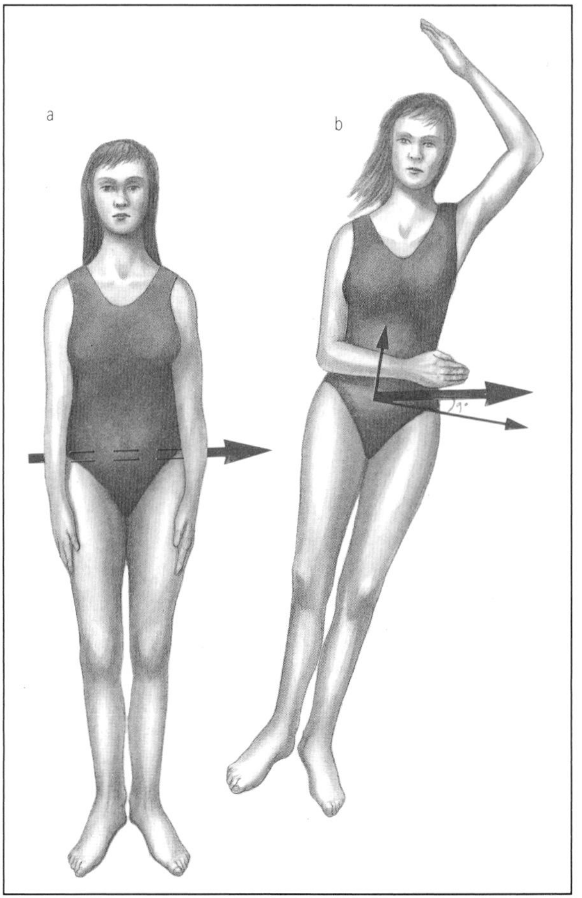

FIG. 7.16 *Diagrams to explain how the twist is started in* FIG. 7.15b.

FIG. 7.16b we can think of it as having a somersaulting component and a twisting component. If the body tilts through 9 degrees, the twisting component of angular momentum is only 0.16 of the somersaulting component. However, the moment of inertia about the twisting axis is only about 1/15 of the moment of inertia about the somersaulting axis, so to have 0.16 times as much angular momentum the twist must have 15 × 0.16 = 2.4 times the angular velocity: the body will make 2.4 twists for every somersault.

Twists can also be started in mid air by other movements. I chose the example of raising an arm because it is conveniently simple.

CHAPTER EIGHT

CYCLING

People can cycle very much faster than they can run. For example, the world record time for cycling ten kilometres (11 minutes 39.7 seconds) is less than the record for running only five (12 minutes 58.4 seconds): the corresponding speeds are 14.3 and 6.4 metres per second. Again, the world record speed for cycling 200 metres with a flying start is 19.8 metres per second, but the peak speed in the 100 metres final of the 1988 Olympic Games was only 12.0 metres per second. (Ben Johnson and Carl Lewis both reached this speed).

Cycling is faster because it needs less power, for the same speed. We know this from many experiments, in which rates of oxygen consumption have been measured. In one particular experiment, cyclists on racing bicycles rode up and down the runway of a disused airfield, with a car driving alongside. The air that they breathed out was sucked from the face masks that they wore, to analysis equipment in the car, through vacuum cleaner tubing. Data from such experiments are shown in FIG. 8.1. The speeds at which the measurements were taken were low enough for the athletes to maintain oxygen balance, so metabolic power could be calculated from oxygen consumption. The resting metabolic rate has been subtracted. These graphs show that for any speed at which people can run, cycling needs much less energy. However, cycling and running at the record speeds (given above) for 5 or 10 kilometres both require power at the rate of about 2000–2500 watts. Speed skating at the 10 kilometre world record speed (12.1 metres per second) also needs about this much power.

The rates at which muscles use energy depend on the forces they have to exert, as I explained in Chapter 4. We already know from force plate records that the forces on our feet in running are predominantly vertical, with peaks close to three times body weight (FIG. 8.3a). The forces involved in cycling have been measured by using bicycles with instrumented pedals. The bicycle was kept stationary by riding it on rollers, so as to keep it close to recording equipment to which transducers in the pedals were connected by wires. The records show much smaller peak forces than in running. These act (as you might expect) roughly at right angles to the pedal surface. When the record shown in FIG. 8.3b was made, with peak loads of less than half body weight, the cyclist riding on rollers was exerting as

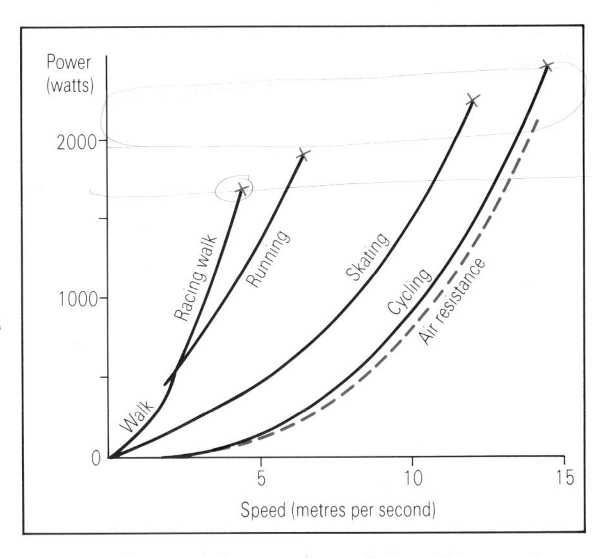

FIG. 8.1 *The metabolic power required for walking, running, skating and cycling, plotted against speed. The graphs have been extrapolated to the world record speeds for 5 kilometre (running) or 10 kilometre races. The broken line represents power required to overcome air resistance. Based on a graph by P.E. di Prampero (1986)* Int. J. Sports Med. **7**, *55–72.*

FIG. 8.2 *Racing cyclists. © Colorsport.*

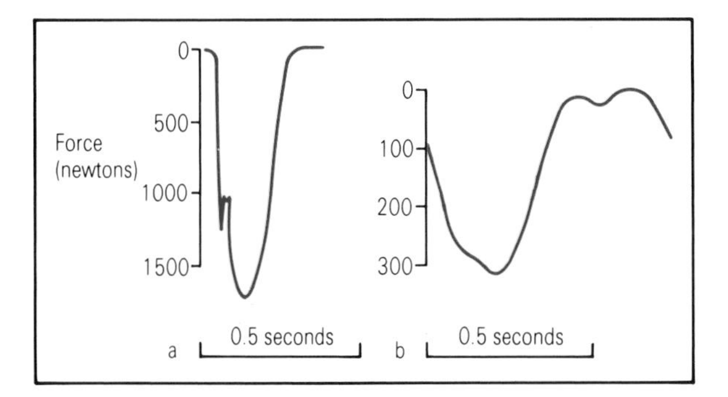

FIG. 8.3 *(a) The vertical force exerted on the ground by one foot of a 70 kg man, running at 3.6 metres per second. (b) The force perpendicular to the pedal surface exerted by one foot of a man cycling on rollers.*
Based on force records by R.McN Alexander and A.S. Jayes (1981) J. Biomechan **13**, *383–390 and by R.R. Davis & M.L. Hull (1981)* J. Biomechan **14**, *857–872.*

much power as in normal cycling at 12 metres per second. (The power was calculated from the forces on the pedals and their movements.)

We should perhaps expect metabolic power to be proportional to the average forces that the muscles have to exert, rather than the peaks. In running the average force on each foot, over a complete stride, is of course half body weight. In cycling the feet do not have to support body weight and the average force can be less: it is about 0.2 times body weight in the example of FIG. 8.3b.

If that were the whole of the story we would expect skating to need as much power as running because the average force on each foot would then also be half body weight. It actually needs much less power than running at the same speed, as FIG. 8.1 shows. This is probably mainly because the leg movements are slower, so slower, more economical muscle fibres can be used. Similarly, fast running requires the same average force as slow running, but needs more power because faster muscle fibres have to be used.

A lot of research on human exercise physiology has used bicycle ergometers, sophisticated exercise bicycles used as scientific instruments. The subject of the experiment sits in the seat and pedals but does not get anywhere: he simply makes a wheel revolve against a brake. This can be tightened to make him work harder or slackened to give him an easier ride, and, because he is stationary, it is easy to connect him to an oxygen analyser or other physiological apparatus.

FIG. 8.4 shows how the stationary cyclist's power output can be measured. This is a highly simplified diagram; in a modern ergometer the same effect would be achieved by more sophisticated means. The brake is a belt rubbing on the rotating wheel. It is kept taut in this simple diagram by a weight which can be increased or reduced to give stronger or weaker braking. If there were no friction between belt and wheel, the force registered by the spring balance would be equal to this weight. Because there is friction it is less than the weight, and the difference is the retarding force on the edge of the wheel. The mechanical power needed to drive the wheel is this retarding force multiplied by the speed of the rim. (The work done by a force is calculated by multiplying it by the distance it moves its point of application, so the power, the rate of doing work, is the force multiplied by the speed.) If as well as the spring balance we have a device to measure the rate of revolution of the wheel we will have the information needed to calculate power.

A person uses oxygen slowly while resting, but uses it faster when pedalling a bicycle ergometer.

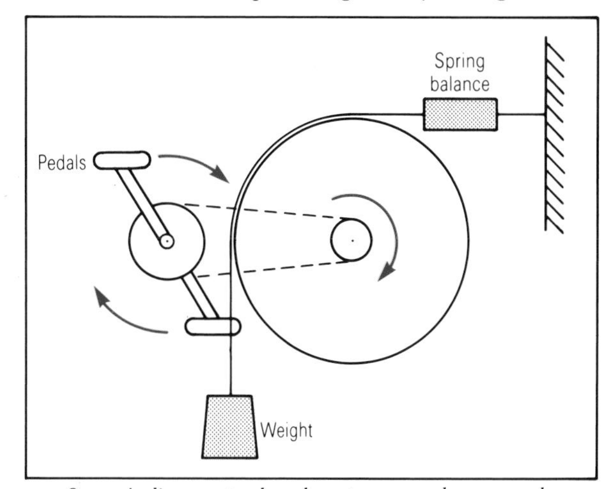

FIG. 8.4 *A diagram to show how power can be measured on a bicycle ergometer.*

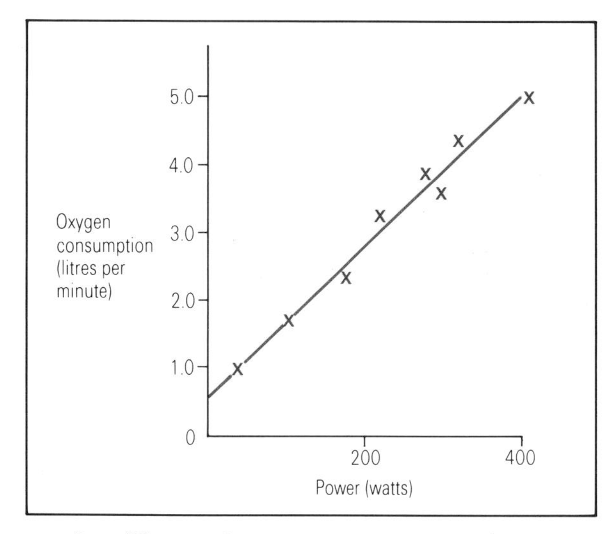

FIG. 8.5 *The rate of oxygen consumption of a professional racing cyclist on a bicycle ergometer, plotted against his mechanical power output. Re-drawn from L.G.C.E. Pugh (1974)* J. Physiol. **241**, *795–808.*

Experiments show that the extra rate is proportional to the mechanical power needed to drive the wheel (FIG. 8.5).

Metabolism using one cubic centimetre of oxygen releases 20 joules of energy from foodstuffs. Ideally that should enable us to do 20 joules of work but life is far from ideal. Ergometer experiments show that even in the best conditions, no more than 5 joules of work are done for each cubic centimetre of oxygen that is used. Twenty joules of food energy yields only 5 joules of work (the rest is lost as heat) so the efficiency of the process is 5/20 or 25%.

A cyclist can exert the same power by pedalling fast against a small braking force or slowly against a large one. Men who are allowed a choice pedal at about 1.3 revolutions per second when high power output is needed, or rather less at low power output, and get the optimum efficiency of 25%. Faster and slower pedalling rates are less efficient but the effect is not very marked. At fairly high power outputs the efficiency is 22% or more (very close to the optimum) at all pedalling rates between about 0.6 and 1.8 revolutions per second.

When we discussed the metabolic energy cost of walking and running I emphasized the forces that muscles have to exert: metabolic energy is used whenever a muscle exerts force. In this chapter, writing about cycling, I am emphasizing the work they have to do: muscles use metabolic energy when they do work. The reason for the apparent inconsistency is that there is a very important difference between going on foot and on a bicycle. In walking and running the muscles get stretched, acting as brakes, and then shorten, doing work. They do relatively little net work: the work done when they shorten is only slightly more than the energy lost in them as heat in their braking phase. While being stretched they use metabolic energy more slowly and while shortening they use it faster but on average they may use it at about the same rate as if they kept constant length while exerting the same forces. In cycling, however, there is very little braking by muscles: they shorten, doing work, during almost all of the time they are exerting forces. (I will soon show why that work is needed.) They use much more metabolic energy than if they kept constant length. It is better to think of their task as doing work, rather than merely exerting force. Cycling is not as much more economical than running as my remarks about the mean forces may have suggested.

Ergometer experiments make it seem likely that cyclists generally work with efficiencies of about 25%, using four joules of metabolic energy for every joule of mechanical work that they do. That means that we can estimate the mechanical power output of cyclists by dividing the metabolic power (measured in experiments such as the one on the airfield) by four. Divide this mechanical power (the rate of doing work) by the speed and you get the work needed to travel unit distance. FIG. 8.6 shows work per unit distance plotted against speed squared. This graph is a neat straight line: we will soon see why.

Some work has to be done against friction in the mechanism of a bicycle but if the machine is a good one, kept well lubricated, that work is small enough to ignore. Work is also needed to overcome the 'rolling resistance' of the wheels, which needs explanation. As the wheels rotate, different parts of their tyres take the weight and are distorted. They recoil to their original shape when the load is taken off again. Just as some of the work done stretching a tendon is lost as heat instead of being returned in its elastic recoil, so also some of the work done on a tyre gets lost as heat. (And tyres get warm as they travel.) Also, the ground that the wheel crosses is distorted by the load and recoils and again, some energy is lost as heat. Whatever the speed, the wheels turn the same number of times to travel unit distance, so the work per unit distance done against rolling resistance is about the same at all speeds. It is smaller for tyres that are inflated hard than for soft ones and it is also smaller for cycling on smooth concrete than on soft mud.

Work also has to be done against aerodynamic drag, as is obvious if you cycle against the wind. Drag acts on any object moving through air (or through water or any other fluid). It depends on the size, shape and speed of the object and on the density of the fluid, and is calculated from the equation

$$\text{drag} = \tfrac{1}{2}\,(\text{fluid density}) \times (\text{frontal area}) \times (\text{speed})^2 \times \text{drag coefficient}.$$

The frontal area is the area of a front view of the object, looked at along the direction of motion (FIG. 8.7). The drag coefficient is a number that depends mainly on the object's shape.

When spheres, cylinders, people and other unstreamlined objects move through air, they leave a large wake of swirling air behind them. Work must be done as the object passes, giving kinetic energy to this air, so relatively large forces are needed to pull it through the air. The drag coefficients of such objects are generally between 0.5 and 1.0. Streamlined bodies are shaped to disturb the air as little as possible. (The best shapes are torpedo-like, rounded in front and tapering to a point behind. They have much lower drag coefficients, generally about 0.05.) The drag coefficients given in this paragraph are about right for objects of a wide range of sizes and speeds (including the ones that interest us in the context of cycling) but do not apply to very small or very slow objects such as airborne dust particles.

The work done against drag is the drag multiplied by the distance travelled, so the work per unit distance equals the drag. We have seen that it is proportional to the square of the speed but that the work per unit distance done against rolling resistance is constant, irrespective of speed. The straight line in FIG. 8.6 tells us that work per unit distance (W) is related to speed (v) by an equation like

$$W = A + Bv^2$$

where A and B are constants. The obvious inference is that A is the rolling resistance and Bv^2 is the drag. This has been checked by suspending cyclists in a large wind tunnel and measuring the drag on them, and seems to be correct.

The line in FIG. 8.6 cuts the vertical axis at about seven joules per metre, so that must be the value of A, the work per unit distance used to overcome rolling resistance. It is a large fraction of the work needed for cycling slowly but a very small proportion of the work needed for cycling fast. Remember that the cyclist was riding over the smooth tarmac of an airfield: rolling resistance would have needed much more work at all speeds if he had been riding through thick mud.

Runners, skaters and cyclists adopt different postures. Cyclists have to overcome drag on the bicycle

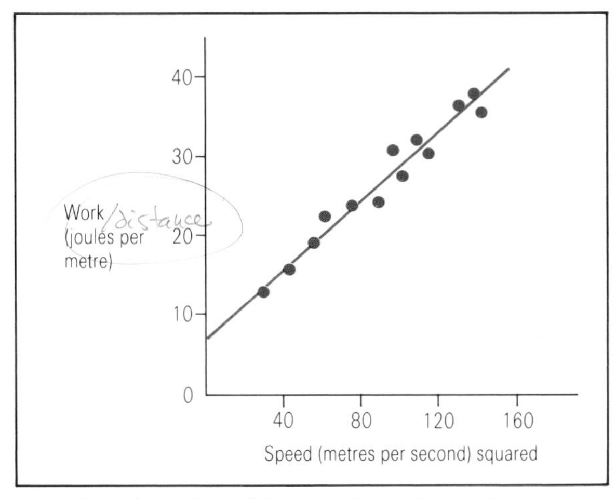

FIG. 8.6 *The work needed to travel unit distance by the same cyclist as in* FIG. 8.5, *plotted against the square of the speed. Re-drawn from L.G.C.E. Pugh (1974)* J. Physiol. **241**, *795–808.*

FIG. 8.7 *A frontal view of a cyclist. The frontal area is the area of the piece of paper you would get if you enlarged this picture to life size and cut round the outline of bicycle and rider.* © *Colorsport.*

FIG. 8.8 *The Vector Single. Photo supplied by Du Pont de Nemours International SA, Geneva.*

as well as on their bodies. The drag at any particular speed is nevertheless about the same, for all three modes of travel. The broken line in FIG. 8.1 shows the metabolic power needed to overcome drag. (It was calculated by multiplying drag by speed to get mechanical power, and then multiplying by four to take account of the efficiency of the muscles.) Notice that it is a large proportion of the power needed for cycling at all but the slowest speeds, but only a small fraction of the power needed for running.

The rules of cycle racing allow competitors to wear tight-fitting clothes to reduce drag, but do not allow streamlining. If they did, much higher speeds would be possible. The Vector Single (FIG. 8.8) is a sophisticated streamlined pedal car. It is wider than a cyclist's body but very much lower, and has about the same frontal area as a racing cyclist on his bicycle. Its streamlining gives it a drag coefficient of only 0.11, compared to 0.75 for a racing cyclist, so the drag on it is only about one seventh as much as on a racing cyclist at the same speed. It has done a flying 200 metres at 26.3 metres per second (59 miles per hour), considerably better than the bicycle record of 19.8 metres per second. A two-man version has gone even faster.

Drag can be greatly reduced in cycling by riding close behind another cyclist or (dangerously!) a motor vehicle. Groups of cyclists can travel faster than individuals if the members take turns to go in front, shielding the others from drag. A cyclist riding close behind a specially prepared car has reached the remarkable speed of 68 metres per second (152 miles per hour).

Many of the current world cycling records were made at Mexico City (altitude 2230 metres) or La Paz, Bolivia (3600 metres). There is less drag at high altitudes than at sea level, because the air is less dense, but it would be a mistake to try to break records at too high an altitude because of the difficulty of getting enough oxygen to breathe.

The work done by a shortening muscle is the force it exerts multiplied by the distance it shortens. The power output, the rate of doing work, is the force multiplied by the rate of shortening. FIG. 8.9 shows (like FIG. 3.3) that the faster a muscle is shortening, the less force it can exert. If it shortens at zero rate (that is, if it keeps constant length) it can exert a large force, but the power is zero. If it shortens at its maximum rate it exerts no force so the power is again zero. It exerts maximum power at an inter-

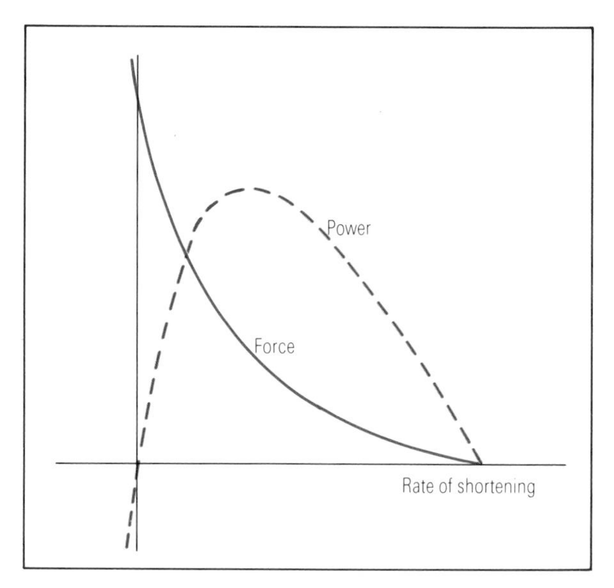

FIG. 8.9 *The force that a muscle can exert, and the corresponding power output, plotted against the rate of shortening.*

mediate speed, usually about one third of the maximum. Experiments with athletes on a bicycle ergometer showed that they could produce most power neither when pedalling very fast against a small load, nor when pedalling slowly against a very large one. They produced maximum power only when pedalling moderately fast against a moderate load. Cyclists use different gears when sprinting uphill than on level ground, to keep their muscles working near the optimum for power output. In longer races, muscle efficiency (the ratio of work done to metabolic energy used) may be more important, but that too is greatest at moderate pedalling speeds.

Bicycle ergometers have sometimes been used to measure the maximum power output that people can produce in a brief spurt of activity. They measure only power from the leg muscles, but a large proportion of the body's muscle is in the legs. In one set of tests, ten male sprinters (eight of them members of the French national track and field team) produced maximum powers of (on average) 1221 watts (1.64 horsepower), in bursts of about six seconds. A group of rugby forwards produced the same mean maximum, but they were heavier, 91 kilogrammes compared to 72. The power per unit body mass was 17 watts per kilogramme for the sprinters and 13 watts per kilogramme for the forwards. These and other data are shown in TABLE 8.1. They show that the sprinters produced more power for their mass than the other athletes: that the endurance runners were no more powerful than a group of young men who exercised only for recreation: and that the women were less powerful than men at the same level of athletics, even when allowance is made for their smaller body masses.

A different picture emerges when sustained power output is examined. In another investigation, athletes from the University of Tennessee track team pedalled a bicycle ergometer while their rates of oxygen consumption were measured. The load that they had to pedal against was increased at two minute intervals until they could not continue. The highest rate of oxygen uptake during this test was 4.0 litres per minute for the five distance runners in the group but only 3.6 litres per minute for the sprinters. The contrast was more striking if allowance was made for the difference in body mass (65 and 77 kilogrammes, respectively): the distance runners used 0.061 litres

TABLE 8.1 *Maximum power output in bursts of about six seconds on a bicycle ergometer, from the data of H. Vandewalle* et.al. *(1987)* Eur. J. appl. Physiol. **56**: *650–656.*

		Number of subjects	Mean mass, kg	Mean maximum power W	W/kg
Male	sprinters	10*	72	1221	17.1
	rugby forwards	17	91	1221	13.3
	hockey players	12*	68	1011	14.8
	endurance runners	12	64	758	11.7
	recreational	7	65	813	12.4
Women	sprinters	5*	56	796	14.1
	recreational	7	59	594	10.2

*Most or all members of these groups were members of the appropriate French national team.

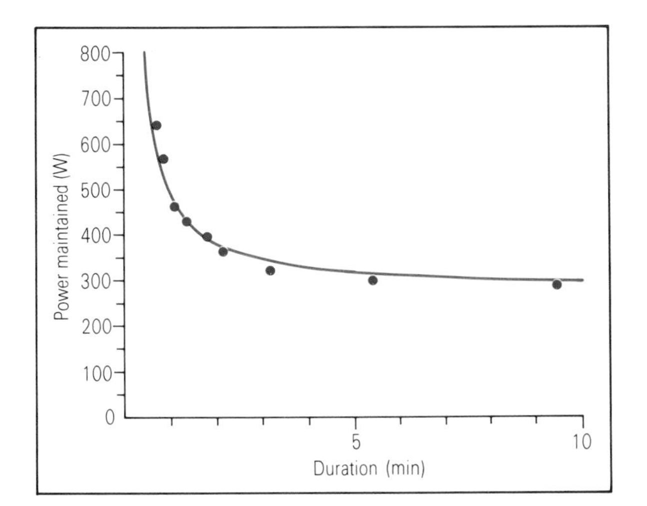

FIG. 8.10 *The mechanical power output that an athlete on a bicycle ergometer was able to sustain for different times, from D.R. Wilkie (1985)* J. exp. Biol. **115**: *1–13*.

per kilogramme per minute and the sprinters 0.047. The distance runners were also more efficient than the sprinters, using a little less oxygen at any particular mechanical power output. Since they could burn fuel faster and were more efficient they could presumably sustain higher power outputs than the sprinters, but this information was not given in the paper describing the experiments.

High power outputs like those given in TABLE 8.1 can be maintained only for a few seconds. FIG. 8.10 shows that an athlete who could sustain 650 watts for 40 seconds on a bicycle ergometer (and could presumably have managed more in a shorter burst of activity) could sustain only 300 watts for ten minutes.

FIG. 8.11 shows the most remarkable pedal-powered vehicle ever built, a single seat aircraft that was pedalled across the English Channel in 1979. It was exceedingly light (only 27 kilogrammes) and correspondingly fragile. Its enormous wings enabled it to fly very slowly, and 250 watts was enough power to keep it airborne, but the pilot had to sustain this power for three hours.

FIG. 8.11 *The Gossamer Albatross. Unlike conventional aircraft, it has the elevator in front and the propeller behind. © The Hulton Picture Company.*

CHAPTER NINE

SWIMMING

People are slow swimmers. The world record times for 50 metres freestyle represent speeds of 2.29 metres per second for men and 2.00 metres per second for women, far below the top speeds of large fishes and whales. In an idyllic experiment, scientists trained a tropical spotted dolphin to swim fast across a lagoon in Hawaii. They trained it to follow a lure towed at variable speed by a winch, and recorded brief bursts of swimming at up to 11 metres per second. That equals the peak speed of good sprinters in a 100 metre race, whereas the human speed record in water would be only a fast walking speed on land. (The dolphin of course cannot walk or run: it is superior only in its own medium.)

People have very nearly the same density as fresh water, 1000 kilogrammes per cubic metre. Muscle is slightly denser than water, about 1060 kilogrammes per cubic metre (drop a chunk of meat into water and it will sink). Most of the other soft tissues have about the same density but bone is about twice as dense as water and fat slightly less dense than water. The density of the body as a whole depends very much on the air in our lungs: measurements on 27 young women showed that their average densities were 980 kilogrammes per cubic metre with lungs fully inflated and 1020 kilogrammes per cubic metre with them deflated. However, there was a good deal of individual variation, and five of the women (fatter than the others) were less dense than water even after breathing out as completely as they could. Sea water is denser than fresh, about 1026 kilogrammes per cubic metre, so it is easier to float in the sea.

Think of a 50 kilogramme woman of density 1000 kilogrammes per cubic metre (her lungs are probably half inflated) floating in seawater. Archimedes' principle says that the water will exert an upward force on her body, equal to the weight of water that she displaces. That means, the weight of a volume of water equal to the submerged volume of her body. If this is to support her weight and prevent her from sinking, the mass of water that she displaces must be 50 kilogrammes. If the density of this water is 1026 kilogrammes per cubic metre its volume is 50/1026 = 0.0487 cubic metres. The volume of her entire body is 50/1000 = 0.0500 cubic metres. Thus 97.5% of her body must be submerged and only 2.5% (less than half the volume of her head) above the surface. For her to be able to breathe, that 2.5% must include her mouth and nose. Conveniently, we tend to float head up because our lungs, the least dense organs in the body, are near the head end.

To propel ourselves forward through water we must push water backwards. How much must we push, and how fast? Newton's second law of motion says that the force needed to accelerate a body is the body's mass multiplied by its acceleration. Momentum is mass multiplied by velocity and acceleration is rate of change of velocity, so we can express the law in a different way

$$\text{force} = \text{rate of change of momentum.}$$

Suppose that in unit time I accelerate mass m of water to velocity v. I am giving momentum to the water at a rate mv and producing thrust mv to propel myself. I am also giving kinetic energy to the water at a rate $\frac{1}{2}mv^2$, and my muscles must do work to supply this energy.

Just as aerodynamic drag resists movement through air, hydrodynamic drag resists movement through water. The thrust in swimming is needed to overcome the drag, so the drag must equal the thrust mv. If the swimming speed is V, the rate of doing

FIG. 9.1 *Stages of the crawl. After J.E. Counsilman (1968)* The Science of Swimming *Pelham Books, London.*

work against the drag (the force multiplied by the velocity) is mvV. Think of this as the useful work that *must* be done to drive the body through the water. In addition, work is done at a rate $\frac{1}{2}mv^2$ giving kinetic energy to the water so the total power output must be $mvV + \frac{1}{2}mv^2$ and we can calculate the propeller efficiency, the useful power divided by the total mechanical power.

$$\text{propeller efficiency} = \frac{mvV}{mvV + \frac{1}{2}mv^2} = \frac{V}{V + \frac{1}{2}v}$$

The smaller the velocity v that the swimmer gives to the water, the more efficient swimming will be. If v is small m must be large to give the required thrust mv. It is more efficient to get your thrust by accelerating a lot of water to a low speed than a little to a high speed, but our small hands and feet cannot push on large masses of water. We are at a disadvantage compared to dolphins which have large tail flukes. Flippers, however, are worn by skindivers and enable them to swim more efficiently.

The crawl (FIG. 9.1) is the fastest swimming stroke, so is used in freestyle races. The arms sweep through the water in turn and the legs kick up and down. Some swimmers kick each leg up and down three times in each cycle of arm movements and some only once. The arm movements seem to be much more important than the leg movements. In one investigation, swimmers were towed through the water by a rope connected to a winch. The force needed to pull them was measured while they remained passive or kicked their legs as in a crawl. Below 1.5 metres per second the kicks gave some thrust, reducing the towing force, but at higher speeds they seemed ineffective. In another experiment swimmers' oxygen consumption was measured while they swam with their arms alone (with a float attached to their feet, to buoy them up), with their legs alone (holding a float in their arms) or with both. They could swim faster with their arms alone than with their legs alone and used less oxygen at any particular speed. Not surprisingly, swimming with both arms and legs was faster and more economical than with either alone. Both the arms and the legs play their part in the crawl, but the arms are the more important.

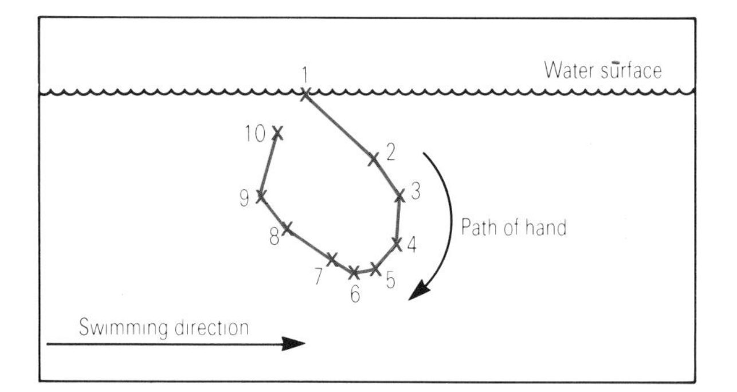

FIG. 9.2 *The path of a swimmer's hand through the water, in the crawl. Based on a film sequence taken through the transparent wall of a pool, reproduced by J.E. Counsilman (1968)* The Science of Swimming *Pelham Books, London.*

FIG. 9.2 shows the path through the water of the hand of a world-class swimmer, doing the crawl. Between stages 4 and 8 it is moving more or less horizontally, moving backwards at about the same speed as the body is moving forwards. Water is being pushed backwards at the speed of the hand, suggesting that v is about equal to V, making the propeller efficiency $V/(V + \frac{1}{2}v)$ about 0.67. This is only a very rough estimate since the forearms and feet may have provided thrust, as well as the hands, but it lies within the range of values that have been obtained by more sophisticated methods.

During stages 4 to 8 (FIG. 9.2) the hand is moving backwards and must work like an oar: because it moves backwards the drag on it acts forwards and serves to propel the body. However, in stages 2 to 4 it is moving almost vertically downwards. Drag on it then acts upward and has no propulsive effect. That does not mean that it cannot be providing thrust; in those stages it may work like a blade of a ship's propeller, producing thrust at right angles to its direction of movement. Forces like this are called lift because the weight of an aeroplane is supported by them. They act whenever a flattish body moving through water or air is tilted at an angle to its direction of movement (FIG. 9.3). Crawl swimmers tilt their hands in the appropriate way in stages 2 to 4 of the stroke, but we do not know how much lift acts on them then or how important it is for swimming. Our hands are very badly shaped for producing lift and would work much better if they were like dolphin tail flukes, with a wide span and a streamlined section.

The force needed to tow an adult man at the surface of water (as if he were swimming) is about 25 newtons at 1 metre per second (a moderate swimming speed) and 100 newtons at a sprint speed of 2 metres per second. These are about nine times the forces needed to tow a man-sized dolphin at the same speeds (calculated from film showing how dolphins slow down when gliding with their bodies rigid, after a burst of speed). The difference is partly due to people swimming at the surface, and dolphins deeply submerged: we produce a bow wave and have to push it along in front of us, but they do not (except of course when they come to the surface to breathe). A large part of the difference, however, is due to dolphins being much better streamlined than people. Evolution has designed us principally for life on land, which has very different requirements from swimming.

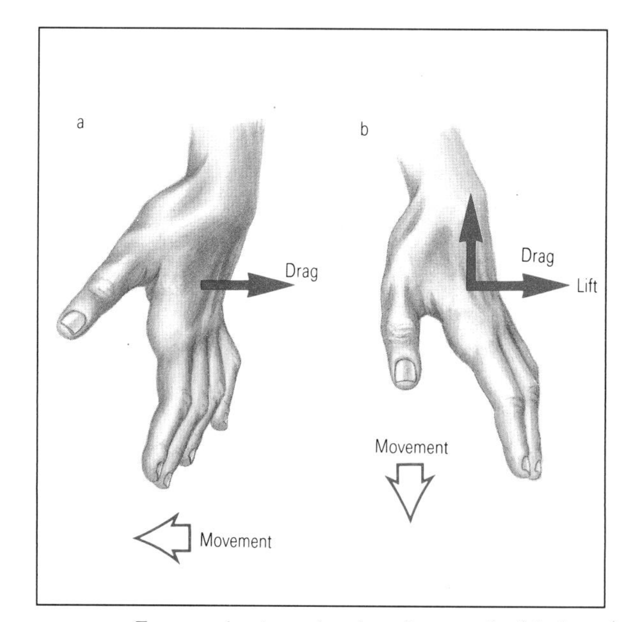

FIG. 9.3 *Forces on hands moving through water. In (a) the hand is set at right angles to its (horizontal) direction of movement and only drag acts on it. In (b) it is tilted at an angle to its (vertical) direction of movement and lift acts as well as drag.*

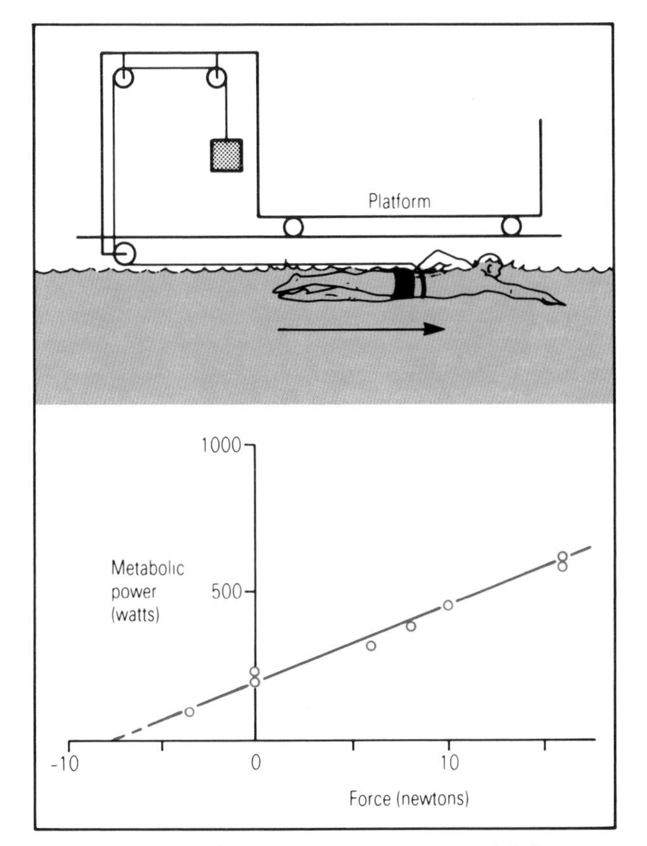

FIG. 9.4 *An experiment to measure the power needed for swimming, and results for a swimming speed of 0.4 metres per second. Metabolic power (above the resting rate) is plotted against the retarding force applied through the rope. After P.E. di Prampero* et al. *(1974)* J. appl. Physiol. **37**, *1–5*.

You might try to calculate the mechanical power needed for swimming from the drag measured by towing. Even if you took account of the propeller efficiency you would get too low a result because the movements of swimming affect the drag. A more reliable method was devised and used by Dr Pietro di Prampero and his colleagues at Buffalo. They used a ring-shaped pool, 2.5 metres wide and deep and 22 metres in diameter, bridged by a platform that could be driven around the ring with the swimmer. A rope connected the swimmer to a weight suspended from pulleys in the cart (FIG. 9.4). This exerted a force on the swimmer that added to the drag or (with the platform reversed, pulling from in front) partly counteracted it. A tube from a face mask carried the air that the swimmer breathed out to oxygen analysis equipment on the platform. The platform was driven at a set speed, the swimmer swam to keep level with it, and oxygen consumption was measured. From this the metabolic power was calculated.

If a swimmer swimming at speed V has to pull against a weight W the mechanical power he has to exert is increased by WV. FIG. 9.4 shows that weights pulling backwards on a swimmer increase his power consumption and weights pulling forward decrease it. At 0.4 metres per second (the speed to which this particular graph refers) it seems that a forward pull of 7.5 newtons would reduce oxygen consumption to the resting rate. This force would reduce the mechanical power requirement by $7.5 \times 0.4 = 3$ watts, so we can estimate the power needed for swimming at 0.4 metres per second as 3 watts. A similar experiment at 1.7 metres per second indicated that the power needed at this higher speed was 210 watts. This huge increase with speed is what we should expect. Drag on poorly-streamlined bodies is proportional to speed squared so power (drag multiplied by speed) is proportional to speed cubed. Thus if 3 watts is needed at 0.4 metres per second we should expect $3 \times (1.7/0.4)^3 = 230$ watts (close to the observed value) to be needed at 1.7 metres per second.

FIG. 9.4 shows that the metabolic power needed for swimming at 0.4 metres per second (without any added load) is 200 watts. We have just calculated that the mechanical power was 3 watts so the efficiency is very low, only 1.5%. The data for faster swimming, at 1.7 metres per second, give an efficiency of 8% but even that is low compared to the 25% that is possible in cycling. Do not confuse these low overall efficiences with the much higher propeller efficiency. The propeller efficiency is the useful work done against drag on the body, divided by the total work done on the water, but the overall efficiency is the useful work divided by the *metabolic* energy consumption.

The position of our lungs near the upper end of the body helps to make us float head up, which may save us from drowning but is a disadvantage in swimming. If our feet sink, tilting our bodies, we present a bigger frontal area to the water than if we were horizontal, and suffer more drag. The principal function of the kick in crawl swimming may be to raise the feet.

FIG. 9.5a shows how the sinking tendency of the feet has been measured. The swimmer is given a tube to breathe through and is strapped to an underwater seesaw which is supported at the foot end by a force transducer. He or she is positioned with the lungs over the pivot so that breathing in or out does not alter the force registered by the transducer. The foot-sinking moment is this force (corrected for the weight of the seesaw) multiplied by its distance from the pivot. Its average values are 14 newton metres for men but only 7 for women. The difference is much too large to be explained simply by men being bigger: it is due to women having more fat to buoy them up, in their thighs and buttocks.

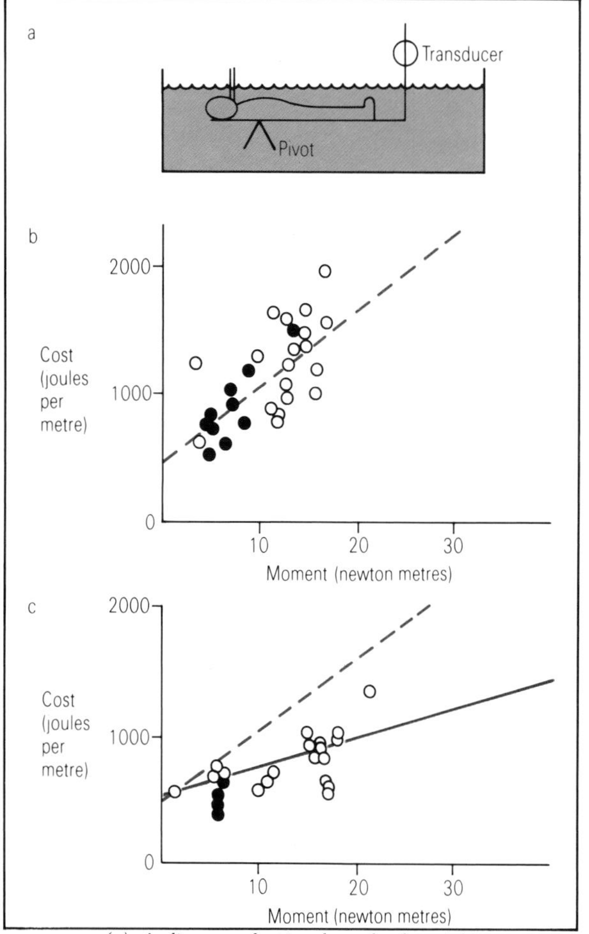

FIG. 9.5 *(a) A diagram showing how the foot-sinking moment was measured and (b,c) graphs of the metabolic energy cost of swimming unit distance plotted against foot-sinking moment for (b) noncompetitive and (c) competitive college swimmers.* ○, *men;* ●, *women. After D.R. Pendergast* et al. *(1977)* J. appl. Physiol. **43**, *475–479.*

The oxygen consumption of the swimmers was measured while they swam fairly slowly, at 0.4 to 1.2 metres per second. For each swimmer, the metabolic rate above resting was roughly proportional to speed, so energy cost per unit distance was constant. The graphs in FIG. 9.5 show that this cost increases as the foot-sinking moment increases, but the effect is much more pronounced for the less skillful swimmers (b) than for the better ones (c). It is less for women than for men (especially than unskilled men) largely because their foot-sinking moments are smaller.

We should remember again that men are generally larger than women. Since drag is proportional to area it seems best to correct for size differences by dividing energy cost by the area of the body surface. This does not alter the conclusion that unskilled women are more economical swimmers than unskilled men. The average costs are 880 joules per metre travelled, per square metre of body surface, for men, and 630 for women.

Among swimmers with large foot-sinking moments, good (competitive) swimmers are very much more economical than poorer (non-competitive) ones (FIG. 9.5). Among those with small moments, however, there is very little difference. It seems that the aspect of swimming skill that has most effect on energy costs is skill in overcoming the sinking tendency of the feet.

FIG. 9.6 compares the energy cost of the crawl with those of other strokes and shows that the crawl is the most economical of energy, at all speeds. It is also the fastest, presumably because it is the most economical: the men's world 100 metre records

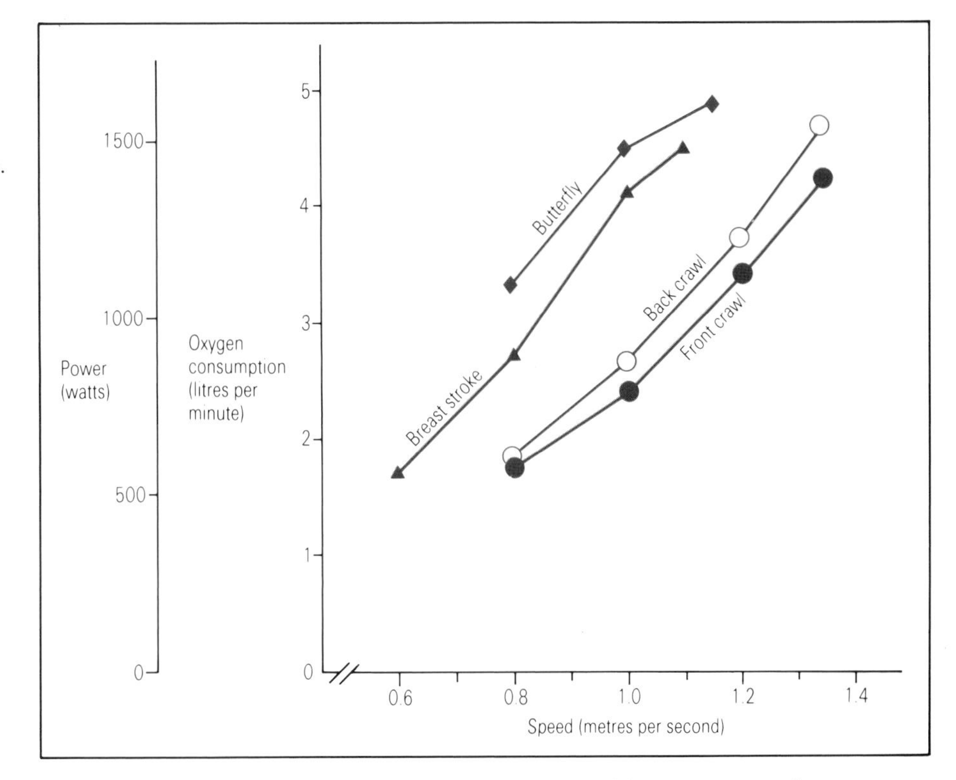

FIG. 9.6 *Rates of oxygen consumption of elite swimmers, using different strokes. After I. Holmér (1979)* Exercise and Sport Sciences Reviews **7**, *87–123*.

are about 48 seconds freestyle (crawl); 53 seconds butterfly; 54 seconds backstroke and 61 seconds breaststroke. If the crawl is so clearly the best, why do swimmers who have the choice ever use the other strokes? The reason may be that poor swimmers with large foot-sinking moments are very much less economical than good ones in the crawl but may be at less disadvantage in other strokes. The energy costs shown in FIG. 9.6 are those of excellent swimmers. Another possible reason is that the crawl involves submerging the face, making skillful control of breathing necessary.

Swimming speeds over long distances, like long distance running speeds, must be limited by the rate at which the heart and lungs can supply oxygen to the muscles. For example, the men's world record freestyle speeds over 50 and 1500 metres are 2.29 and 1.68 metres per second. It was of course not possible to measure Vladimir Salnikov's oxygen consumption when he set the 1500 metre record. If you extrapolated the crawl line in FIG. 9.6 to his record-breaking speed you would get a phenomenally high estimate of oxygen consumption, but this would be misleading: the swimmers of FIG. 9.6 were going continuously around a ring-shaped pool but Salnikov was swimming backwards and forwards along a rectangular one, saving energy by pushing off from the end of the bath after each length.

CHAPTER TEN

THROWING

How far a ball goes depends mainly on how fast you throw it, and at what angle. For any particular angle, the faster the throw the greater the distance. If you throw it vertically it may rise very high but will not travel any distance: it will land at your feet. If you throw it horizontally it will not go far, but you can make it go further by throwing upwards at an angle. The simplest possible theory of throwing (leaving out complications that we will come to later) says that for any particular speed a ball will go furthest if thrown at 45 degrees to the horizontal. This maximum distance is v^2/g, if v is the speed of throwing and g is the acceleration of gravity, about 10 metres per second per second. Thus a throw at 10 metres per second can make a ball travel $10^2/10 = 10$ metres: at 20 metres per second, $20^2/10 = 40$ metres: and at 30 metres per second $30^2/10 = 90$ metres. The distance increases very rapidly as the throwing speed increases.

The Guinness Book of Records says that the fastest recorded baseball pitch was made at 45 metres per second (101 miles per hour: the speed was measured by radar). The simple theory says that speed should send the ball 203 metres, in a 45 degree throw. It was not of course pitched at that angle (it was pitched fast to make it hard to hit, not to send it a long way) but you might suppose it would have been possible for the same pitcher to have thrown the ball 203 metres. The book says however that the furthest anyone is known to have thrown a baseball is only 136 metres. British readers may want to know that baseballs have almost the same mass as cricket balls and that the fastest recorded bowling speed is almost the same as the fastest baseball pitch.

Balls do not travel as far as the simple theory suggests, because they are slowed down by aerodynamic drag. FIG. 10.1 shows that the effect is small for low speeds, but a cricket ball thrown at 30 metres per second goes only two thirds as far as it would go in a vacuum, and the fractional loss is greater at higher speeds. That is in still air: you can throw further with the wind and less far against it. I do not know how fast the wind was blowing when the 136 metre throw was made: the graph shows a distance of only 109 metres for a throw at the record pitching speed of 45 metres per second.

The graphs in FIG. 10.1 have been calculated for throws that start and end at the same level, but balls are generally thrown from a metre or two above the ground. The discrepancy makes little difference in calculations about long throws. If a ball leaves the thrower's hand two metres above the ground, the graph gives the distance to the point at which it is again two metres from the ground. It will be falling at 45 degrees (or a little more steeply if it has been slowed by the air) so will travel two metres further (or a little less), before hitting the ground. An error of a couple of metres is insignificant in most calculations about balls being thrown many tens of metres.

Shot-putting is a throwing competition, but the world record for men is only 23.12 metres, far less than the distance a ball can be thrown. This is not surprising, for the 7.26 kilogramme shot used by men has fifty times the mass of an average baseball. (Women use a 4 kilogramme shot and put it almost as far as the men put their larger one.) Indeed, it may seem surprising that athletes can put the shot so far. For the longest puts, it has to be released at a speed of about 14 metres per second so its kinetic energy (for the men's shot) is $\frac{1}{2} \times 7.26 \times 14^2 = 711$ joules. In contrast, a 145 gramme baseball thrown at

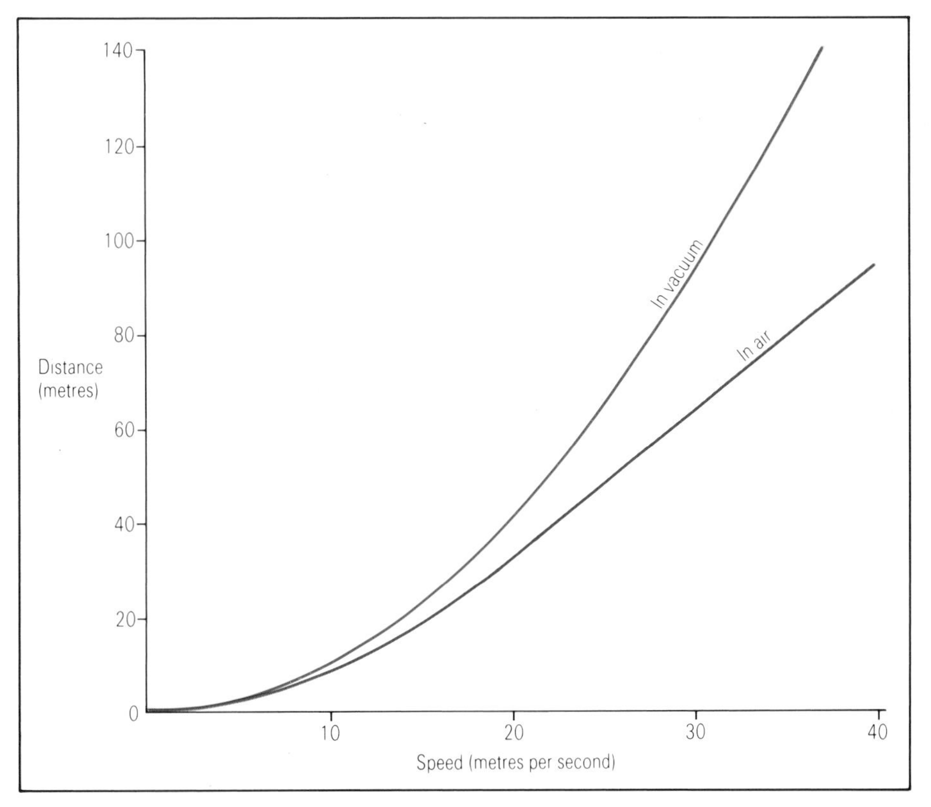

FIG. 10.1 *Graphs of distance travelled against speeds, for cricket balls thrown from ground level at 45 degrees to the horizontal, over level ground. One line shows how far the ball would go in a vacuum, and the other how far in still air.*

45 metres per second has a kinetic energy of only 147 joules. However, much of this discrepancy is explained if you add the kinetic energy that has to be given to the body: the hand (mass about 0.6 kilogrammes) has to be accelerated to the same speed as the ball and the forearm (1.5 kilogrammes) to a slightly lower speed. The energy given to the missile is a much smaller fraction of the total work that has to be done in the case of the light ball, than of the heavy shot. Also muscles cannot exert as much force or do as much work when shortening fast to throw a ball as when shortening more slowly to put a shot.

Shot-putting is done from a 2.1 metre circle. The shot is accelerated to about 3.5 metres per second in the initial glide across the circle (FIG. 10.2b–d), gaining about 44 joules of kinetic energy. The remainder of its kinetic energy, about 670 joules for a really good put, is given to it as it is accelerated over the last 1.7 metres before release (FIG. 10.2, d–f), requiring a mean force (work divided by distance) of almost 400 newtons. Measurements of the acceleration of the shot show that the *peak* force on it is considerably higher, up to at least 600 newtons. This is the force that the hand exerts on the shot. The forces that the feet exert on the ground have been measured by means of force plates and are much higher even than this: for example peak values of 2800 newtons for the right foot and 1800 newtons for the left, of an excellent right-handed male shot-putter. Much larger forces have to be exerted on the ground than on the shot, not merely to support the weight of the body but also to accelerate the body's mass.

Aerodynamic drag has hardly any effect on shot-putting because the speeds are so low (compared to baseball pitching) and the shot so heavy. Drag at any particular speed is proportional to frontal area, and deceleration due to drag is proportional to frontal area/mass which is only about one twentieth as much for a shot as for a baseball.

FIG. 10.2 *Putting the shot. After G. Dyson (1973)* The Mechanics of Athletics *ed 6. University of London Press, London.*

FIG. 10.3 *Throwing the hammer. © Colorsport.*

Good male shot-putters release the shot at a height of about 2.25 metres. Since it travels little more than 20 metres, we have to take account of its initial height in any calculations about its trajectory. It turns out that to make it travel furthest for given initial speed it should be projected not at 45 degrees (as for a baseball) but at about 42 degrees.

The hammer has the same mass as the shot but can be thrown immensely further: the world record is 87 metres, compared to 23. Most of the work of accelerating a shot has to be done by a single contraction of each of the muscles involved, in stages (d) to (f) (FIG. 10.2) but a hammer thrower whirls the hammer around and builds up its kinetic energy gradually, over several revolutions of his body, using many successive contractions of each muscle (FIG. 10.3). Also, because the hammer is at the end of a chain he does not have to accelerate his limbs to its speed. Hammer throwing is limited by the centripetal force that the athlete must exert to prevent the hammer from flying off at a tangent too soon. In particular, he must avoid being lifted off his feet when he swings the hammer high, one revolution before release.

We will return to ball throwing and look at the sequence of movement of the joints. Notice in FIG. 10.4a how the woman's shoulder has made a large part of its movement before her elbow starts to extend, and her wrist moves even later. She is using her arm rather like a whip. When you crack a whip, a wave of bending travels down it, transferring kinetic energy from one part of the whip to another. The light tip of the whip is made to move so much faster than the heavy base that its speed becomes supersonic and it produces a shock wave, the audible crack. The same principle (at subsonic speeds!) seems to be used by skilled throwers.

The sequence of joint movements is not limited to the arms, as the pictures of a baseball player show

FIG. 10.4a *Throwing a ball. After R.L. Wickstrom (1975)* Exercise and Sports Sciences Reviews **3**, *163–192.*

FIG. 10.4b *A skilled infielder throwing a baseball, traced from cine-films taken simultaneously by two cameras. After A.E. Atwater (1979)* Exercise and Sport Sciences Reviews **7**, *43–85*.

(FIG. 10.4b). Movements of the whole body are involved in the throw but most of the rotation of the hips happens between stages 1 and 2, most of the rotation of the shoulders between stages 2 and 3 and the straightening of the elbow between stages 3 and 4. The sequence of joint movement starts far from the ball and moves towards it.

Something similar happens in shot-putting (FIG. 10.5): the right knee, hip, shoulder and wrist reach their peak speeds in turn and the speed of each falls a little as the next joint continues to accelerate.

The analogy to a whip seems convincing. Research workers generally agree that the sequential movement of joints is good, because energy is transmitted from segment to segment of the body and eventually to the ball or shot. However, until very recently there has been no mathematical proof that this style of throwing sends the missile further than it would go if we moved all our joints simultaneously. You may feel satisfied with the observation that athletes find that the sequential style works best. I admit that it does, but still feel the need of a mathematical proof because until I have it I will not be quite sure that the proposed explanation is logically sound. Mathematics is the most rigorous (and therefore the most reliable) form of logic.

As a first rough attempt at a mathematical model I have imagined the man shown in FIG. 10.6 (who looks rather like a shot-putter). To keep the mathematics simple I put all his body mass into his trunk and let him use just two muscles, one at the knee and one at the elbow. I gave those muscles realistic mechanical properties, making them able to exert large forces when shortening slowly, but less when they shortened faster (FIG. 3.3b). I wrote a computer programme that enabled me to follow the course of events, when the muscles were activated simultaneously or in turn. Each muscle shortened at an ever-increasing rate, exerting less and less force, until it could shorten no faster. When the knee muscle could shorten no faster, the foot left the ground and when the elbow muscle could shorten no faster, the missile left the hand. The calculations showed that it would indeed be thrown further if the elbow muscles started contracting some time after

FIG. 10.5 *The speeds of parts of the body during a shot-put, plotted against time. From V.M. Zatsiorsky, G.E. Lanka & A.A. Shalmanov (1981)* Exercise and Sport Sciences Reviews **9**, *353–389*.

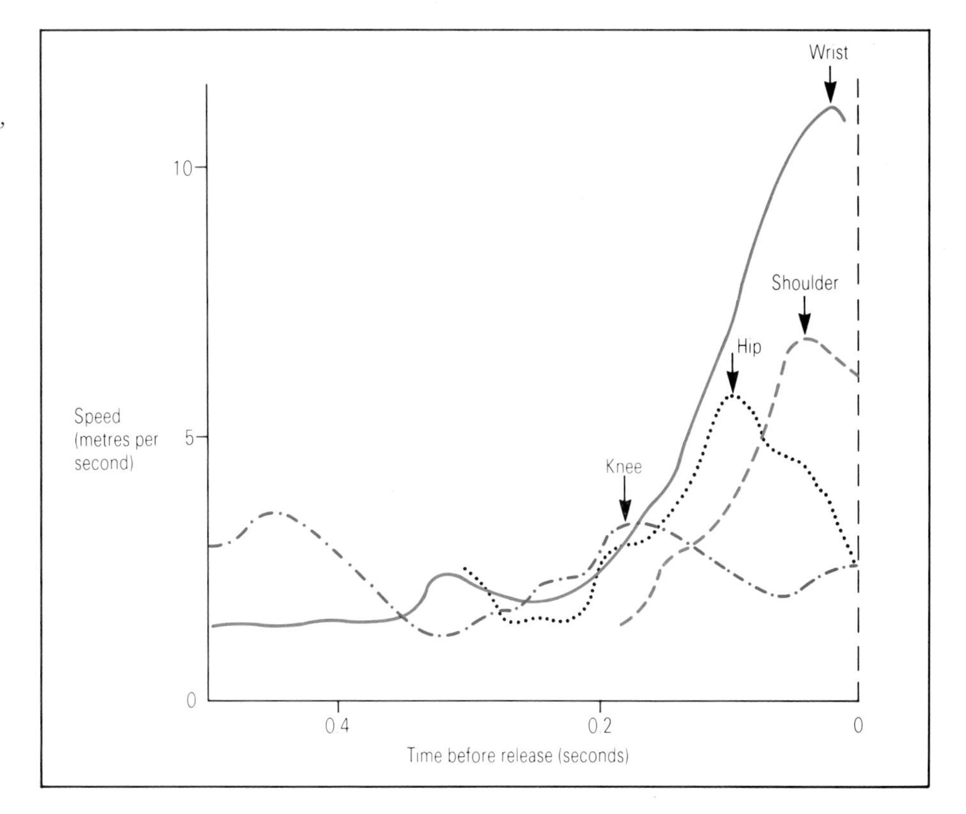

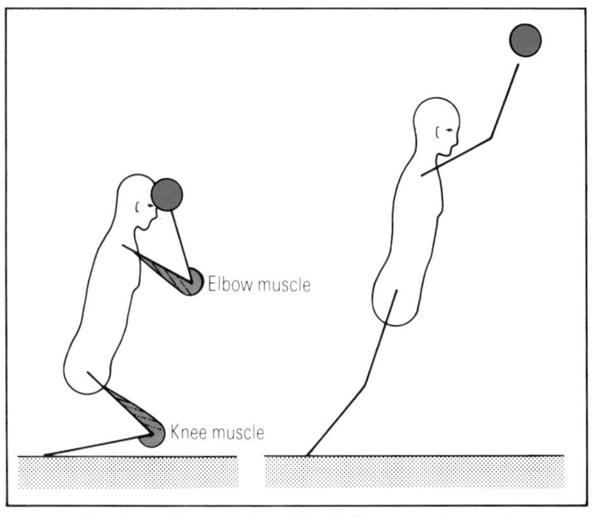

FIG. 10.6 *A mathematical model of throwing, designed to explain the advantage of sequential joint movement.*

the knee muscles. If the elbow muscles were activated too soon, the missile left the hand while the knees were still quite strongly bent:the knee muscle did not have time to shorten much, so did little work. If the elbow muscles were activated too late, they had to shorten faster, so exerting less force and doing less work. This was because the body was by then not accelerating much, so the missile moved faster away from it. There was one particular interval between knee and elbow muscle activation that made the total work (done by the two muscles together) greatest, but it proved better to use a rather longer delay because that increased the proportion of the work that became kinetic energy of the missile. It seems that the whip explanation of sequential movement in throwing is a good deal too simple, and rather misleading.

In some sports, athletes are allowed to run to give extra speed to a throw, but they do not generally choose to run fast. Top javelin throwers (male and

FIG. 10.7 *Successive positions of the feet in the final stride of a javelin throw with arrows representing the forces on the feet. These diagrams, based on films and force plate records of an elite thrower, are after E. Deporte & B. van Gheluwe (1988) in G. de Groot* et al. *(eds)* Biomechanics *XI B, 575–581. Free University Press, Amsterdam.*

female) run up at only about 5.3 metres per second, little more than half what they could achieve in an all-out sprint. This is only a small fraction of the speed of about 30 metres per second at which the javelin leaves the hand, but if they ran faster they might be unable to stop themselves, before crossing the foul line. Again a group of cricketers who were considered to be fast bowlers ran up to bowl at mean speeds of only 4.5 metres per second (when using a 'front on' style of bowling) or 4.0 metres per second ('side on') although they delivered the ball at speeds up to 38 metres per second.

Javelin throwing and cricket bowling involve rather similar foot movements (FIGS. 10.7, 10.8). Force plate records of javelin throwing show that the forces on the feet slope backwards, in the last two footfalls (FIG. 10.7). These forces slow the body as a whole down but may also set it rotating head over heels so as actually to help to increase the speed of the upper parts of the body, and of the javelin. Imagine a wheel thrown so as to fly through the air without spinning, and land on its rim. It would then start rolling, when the part of its rim that was in contact with the ground at any instant would be stationary, but the top part of the rim might be moving forward at a higher speed than the speed at which the wheel was thrown. Formal analysis is needed, to find out whether this principle is actually used. One point that seems to weigh against it is that to set the body rotating in this way, the forces on the feet should not be in line with the body's centre of mass but should point rather behind it. Some of the largest forces shown in FIG. 10.7 seem to point in front of the centre of mass.

A golf ball has about the same ratio of frontal area to mass as a baseball, so could probably be thrown about as far, but can be hit very much further. We must be careful about our comparisons here because

TIA
CUP

the lengths of golf drives, as usually given, include the 'run' (as the ball bounces and rolls along the ground) as well as the 'carry' (the initial flight through the air). It seems not to be unusual for a golf drive to leave the tee at 70 metres per second (156 miles per hour) and the results of tests with a driving machine show that in that case it should carry 220 metres, in still air over level ground. This is fairly normal for a good drive but considerably further than the record-breaking baseball throw of 136 metres. The golf ball goes further, mainly because it leaves the tee very much faster than the baseball leaves the hand, so we need to know why it can be driven so fast.

Part of the reason is that the head of a golf club can be swung faster than the hands can be moved. As the club swings down to hit the ball (FIG. 10.9) the golfer's hands move almost in a circle (in a plane

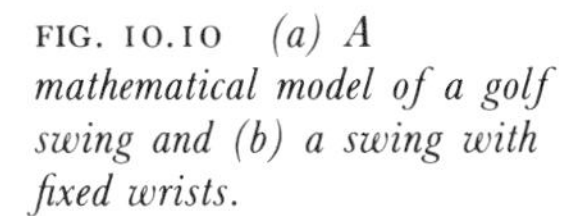
FIG. 10.10 *(a) A mathematical model of a golf swing and (b) a swing with fixed wrists.*

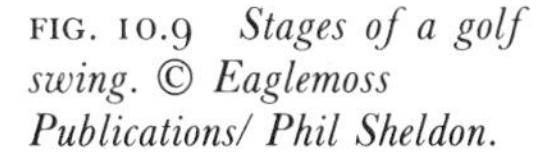
FIG. 10.9 *Stages of a golf swing. © Eaglemoss Publications/ Phil Sheldon.*

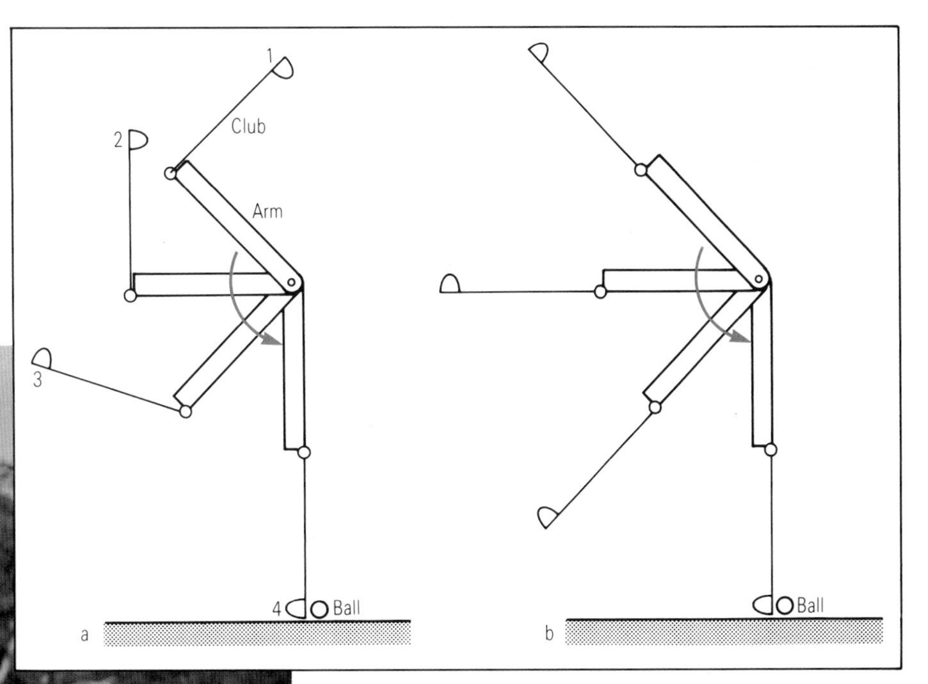

sloping at about 45 degrees to the horizontal), so the swing can be represented by a very simple model (FIG. 10.10a). The arm is represented as a rigid bar, pivoting about a fixed point and moved by strong shoulder and trunk muscles. The club is attached to the arm by a hinge joint at the wrist, which is kept locked at right angles until a stage in the swing when it is suddenly released. Until then, the club head has been moving in a circle, prevented by the force exerted by the wrist muscles from flying off at a tangent (FIG. 10.10a stages 1 and 2). When the wrist is released it is straightened by the club head flying outwards (stages 3 and 4). Ideally, arm and club should be in line with each other when the ball is hit (stage 4) and at this stage the club head is moving very much faster than the hands.

Our wrist muscles are much weaker than our shoulder muscles (they can exert much smaller moments about their joint) so the model can imitate real golf swings quite accurately although it treats the opening of the wrist as passive, ignoring the contribution of the wrist muscles. The club head is accelerated from rest to 50 metres per second in about one fifth of a second, so its acceleration is 250

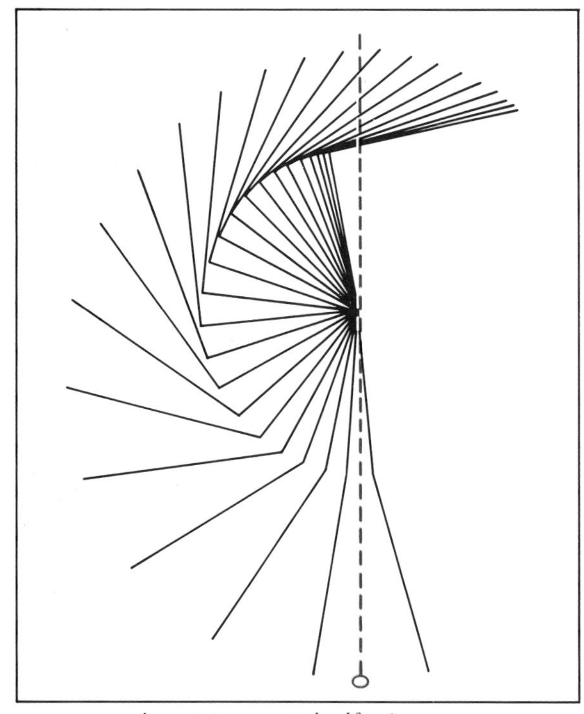

FIG. 10.11 *A computer-generated golf swing.*

metres per second per second, 25 times the acceleration of a body falling freely under gravity. Though the swing is downward and is helped by gravity, this help is so small that we can ignore it in describing the mechanics of the swing.

FIG. 10.11 shows a computer-generated golf swing, calculated for the model that I have just described with the assumption that the shoulders exert constant torque. It is remarkably like a real swing (FIG. 10.9). The distances between successive positions of the wrist and club head show their relative speeds. The wrist actually slows down slightly (retarded by the centrifugal force of the club) as the club head speeds up. At the moment of impact, the club head is travelling about four times as fast as the hands.

It would be possible to swing a golf club without cocking the wrists. If you did that the speeds of wrist and club head would be in proportion to their distances from the centre of the swing: if club and arms had the same relative lengths as in FIG. 10.11, the club head would move only 2.2 times as fast as the wrist. Less of the work done by the shoulder and trunk muscles would have gone into kinetic energy of the club than in the simulation, and more into the kinetic energy of the arms. A rough calculation by C.B. Daish in his excellent book on *The Physics of Ball Games* indicates that 50% of the work done in the swing becomes kinetic energy of the club head, 20% becomes kinetic energy of the shaft and 30% becomes kinetic energy of the body and arms. Just as the sequential movement of the joints in throwing converts as much as possible of the work into kinetic energy of the hand and ball, the delayed uncocking of the wrists in a golf swing gives as much as possible of the energy to the club head. A 0.2 kilogramme club head travelling at 50 metres per second has 250 joules kinetic energy, which is more than can be given to a baseball but less than to a shot.

Did you notice that I gave 50 metres per second as a typical speed for the club head, but 70 metres per second for the ball? That was not a mistake. You can throw only as fast as you can move your hand, but a light ball hit by a heavy club moves faster than the club. High-speed films show that this is true, but to understand why we need to think about force, momentum and energy.

The films show that club head and ball are in contact for only about half a millisecond (1/2000 of a second). The ball is accelerated in this time from rest to 70 metres per second so its acceleration is 140000 metres per second per second. Its mass is 0.046 kilogrammes so the force on it (mass times acceleration) is $0.046 \times 140000 = 6440$ newtons (about two thirds of a tonne). This is the average force during the impact and the peak must be higher, at least one tonne force. This is so much more than the forces that gravity or your hands exert on the club that we can ignore all other forces.

Suppose that a club head of mass M moving with velocity V_0 hits a stationary ball of mass m. After the impact the club has velocity V_1 and the ball v_1. Club and ball exert huge forces on each other but we have seen that we can ignore all other forces, so we can think of club and ball in isolation from any outside influences. That means that their total momentum must remain constant. Before the impact, only the

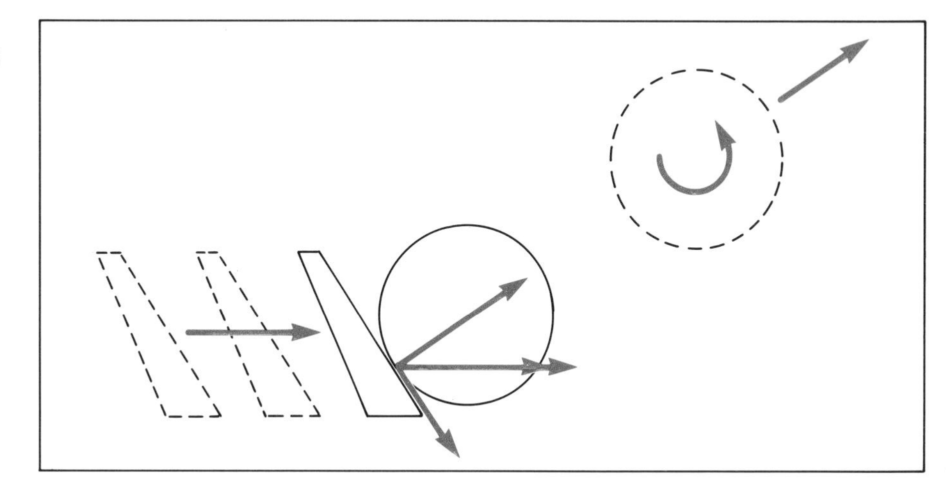

FIG. 10.12 *The tilted face of a golf club makes the ball spin. The arrows represent components of velocity.*

club is moving, with momentum MV_0. After it, the club is moving with momentum MV_1 and the ball with mv_1.

$$MV_0 = MV_1 + mv_1$$

$$V_1 = (MV_0 - mv_1)/M$$

If no energy is lost the total kinetic energy after the impact must be the same as before

$$\tfrac{1}{2}MV_0^2 = \tfrac{1}{2}MV_1^2 + \tfrac{1}{2}mv_1^2$$

If you put the value of V_1 from the previous equation into this one and sort out the algebra you get

$$v_1/V_0 = 2M/(M + m)$$

If the club head and ball had equal masses ($M = m$), the final speed of the ball (v_1) would equal the initial speed of the club (V_0). More realistically, if M is four times m, v_1 is 1.6 times V_0. If M were immensely more than m (if you used a sledge hammer instead of a golf club) v_1 would be almost twice v_0, but that would give you no extra advantage because you could not swing the hammer fast: V_0 would be low. There is an advantage in using a club that is a few times heavier than the ball, but not so heavy as to be cumbersome.

Even the advantage of a club of sensible mass is less than the simple theory suggests because some of the kinetic energy gets converted to heat as the ball deforms and recoils in the impact. The energy losses that occur with real balls reduce the speed of the ball after the impact to about 1.4 times the initial speed of the club head ($v_1/V_0 = 1.4$), but even that is a substantial gain in speed.

Golf shots start and end at ground level so you might suppose that it would be best to hit the ball high into the air, at 45 degrees to the horizontal. Golfers actually hit long shots at much shallower angles, but they carry only a little less far than simple theory (allowing for aerodynamic drag) says a 45 degree shot at the same speed should go. The reason for this is that the ball spins as it flies through the air. I will explain why it spins, and why this affects the carry.

The face of a golf club is tilted back a little, relative to the shaft, so it hits a little below the centre of the ball (FIG. 10.12). The diagram shows that its velocity can be thought of as having two components, one radially towards the centre of the ball (so it pushes the ball in that direction, upwards at an angle) and one tangentially across the surface of the ball (so it makes the ball spin). Double-exposure photographs of marked balls being hit show that drivers (with faces sloping at 10 degrees to the handle) make the ball spin at about 50 revolutions per second. Clubs with faces tilted at larger angles (as for chip shots and for getting out of bunkers) make the ball rise more steeply and spin faster, for example at 130 revolutions per second for a number

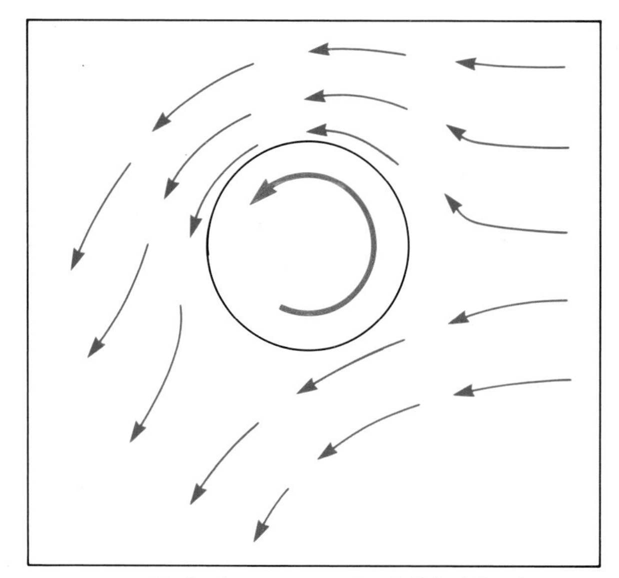

FIG. 10.13 *Air flowing past a rotating ball is deflected as shown in this diagram.*

7 iron. The steeper angle is obtained at the expense of less speed because more of the kinetic energy of the club is used to make the ball spin, becoming kinetic energy of rotation.

It seems easiest to explain the effect of the ball's spin by thinking first about air flowing past a stationary rotating ball (FIG. 10.13). If the ball spins in the direction shown in this diagram it speeds up the air travelling over it and slows down the air travelling under it. The result is that the air is deflected downwards, as the diagram shows (remember that if the wheels on the left side of your car are rotating faster than the wheels on the right, you must be turning right). If the ball is deflecting air downwards it must be pushing the air down, so the air must exert an upward force on the ball: this is called the Magnus effect. Now suppose that instead of having air moving from left to right past a stationary but spinning ball, the spinning ball were moving from right to left (with the same relative velocity) through stationary air. The air movements *relative to the ball* would be the same as before, air would be driven downwards as the ball passed through and the same Magnus force would act.

The direction of spin that the club gives to the golf ball (FIG. 10.12) is called bottom spin. It gives rise to an upward Magnus force, and the effect is enhanced by the dimples on the ball which make its spinning more effective, in moving the nearby air. If the ball is spinning fast enough, as it often is in the early stages of a drive, it may travel in a straight line or even curve slightly upwards, showing that the Magnus force is big enough to counteract its weight. In any case, the Magnus effect delays the ball's fall to the ground, making it travel further. A well-driven ball rises from the tee at a much smaller angle than the 45 degrees that is best for non-spinning balls, but the Magnus effect makes it travel only a little less far than if it were hit at 45 degrees without spin, at the same speed. (To hit it without spin at 45 degrees you would have to drive it from an extraordinarily tall tee, so that the club approached it from below, at 45 degrees).

Top spin, which is often used in tennis shots, gives a downward Magnus force which makes the ball plunge sharply to the ground. Spin also affects the way balls bounce, making spin bowling effective in cricket.

Kicking is a form of hitting and gets the same advantage of speed: if the ball is lighter than the foot it leaves the impact faster than the foot approaches it. For example, in a fast soccer kick the foot approaches the ball at about 20 metres per second and the ball leaves at about 27 metres per second. The movements of the upper and lower leg, as the foot swings towards the ball (FIG. 10.14), are rather like the movements of arm and club in a golf swing (FIG. 10.11). The sequential movement of the joints (hip before knee) ensures that the boot is travelling as fast as possible immediately before impact, just as sequential movement also helps throwing.

FIG. 10.14 *Kicking a ball. © Colorsport.*

JVC
JVC

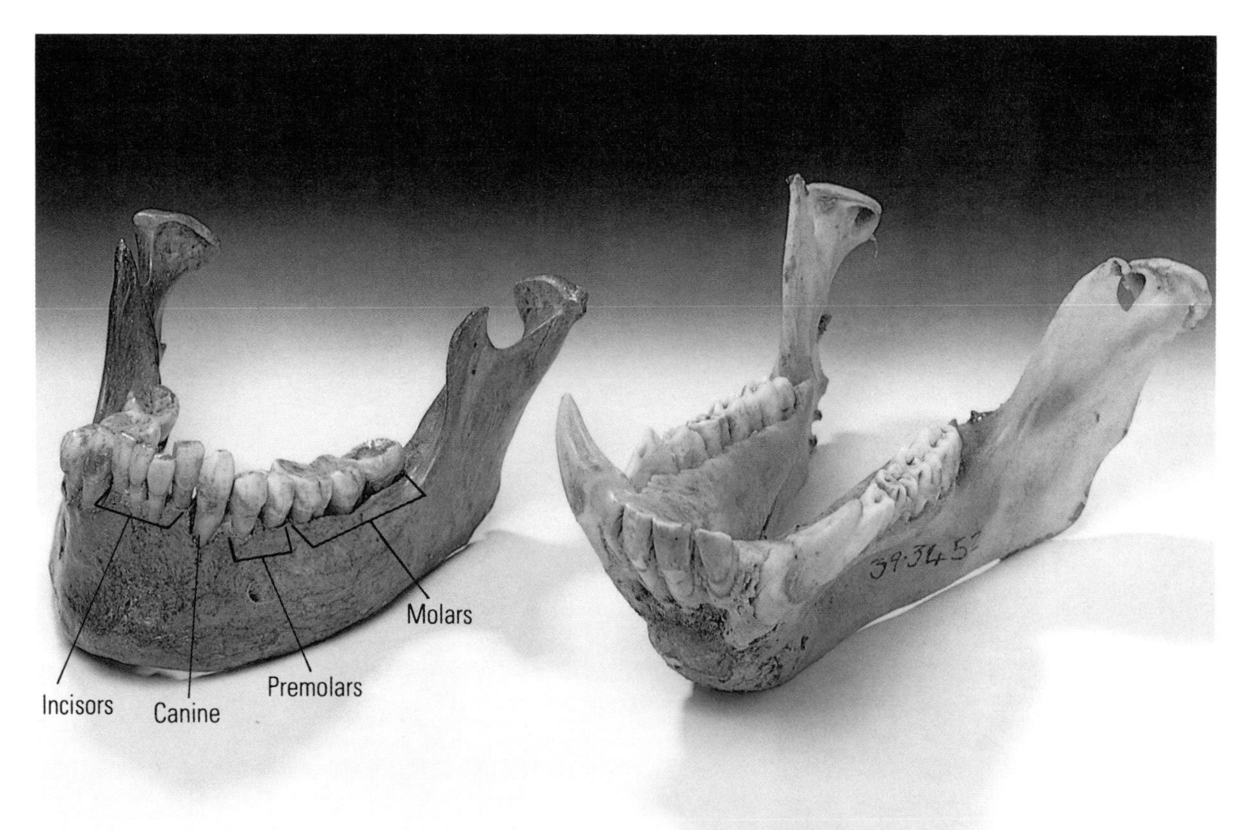

FIG. 11.1 *(right) A section through a tooth.*

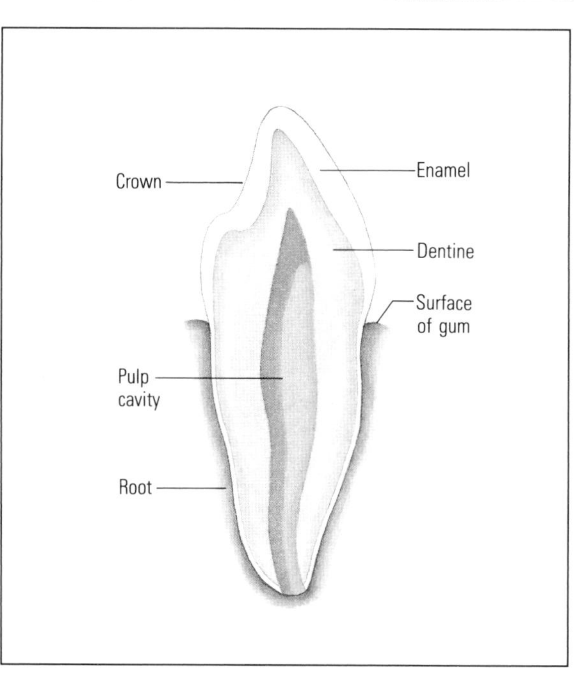

FIG. 11.2 *(above) A human lower jaw, and the lower jaw of a baboon.*

CHAPTER ELEVEN

EATING

This chapter is largely about the mechanics of chewing and swallowing, but also has a little about processes further down the gut.

Teeth are made of dentine (ivory), a material that is very like bone, with an outer covering of enamel which contains more mineral and less protein (FIG. 11.1) Enamel is harder than bone or dentine, so more resistant to wear, but it is brittle and sometimes gets broken. The less-brittle dentine does little to reinforce the enamel against damage, because enamel is much stiffer than dentine (in technical terms, its Young's modulus is three or four times higher). Similarly, if teeth were made of porcelain filled with wood, the less-stiff wood would contribute very little to their strength. The enamel is about 1.5 millimetres thick, much more than enough to make it unlikely to wear through in a lifetime, at least on modern diets. It probably needs to be thick to be strong enough.

An adult with a complete dentition (including wisdom teeth) has 32 teeth, eight in each side of both the upper and the lower jaw (FIG. 11.2). Of these eight, two (at the front) are incisors, one is a canine, two are premolars and three (at the back) are molars. The milk dentition of young children has the same numbers of incisors, canines and premolars, but no molars. Incisors are used for biting off pieces of food, and premolars and molars for chewing. Our canines are reduced remnants of the dagger-like canine teeth that our evolutionary ancestors used as weapons and for tearing flesh. The canines of large carnivores and even of many monkeys (FIG. 11.2) are much larger than ours.

The incisors are square, spade-like blades. When the jaw is closed in its normal resting position with the upper and lower molars in contact (FIG. 11.3a) the incisors overlap, with the upper ones a few millimetres in front of the lower. To make the incisors meet edge to edge so that they can cut like a carpenter's pincers, the jaw must be moved a little forward (FIG. 11.3b). The ligaments of its joint with the skull are loose enough to allow this, and also some side-to-side movement. FIG. 11.3 shows that the condyles (knobs) of the jaw joint fit into hollows on either side of the skull when the molars are being used (a), but move out of the hollows when the jaw moves forward for biting (b). Try biting pieces off foods such as biscuits or apples and notice whether

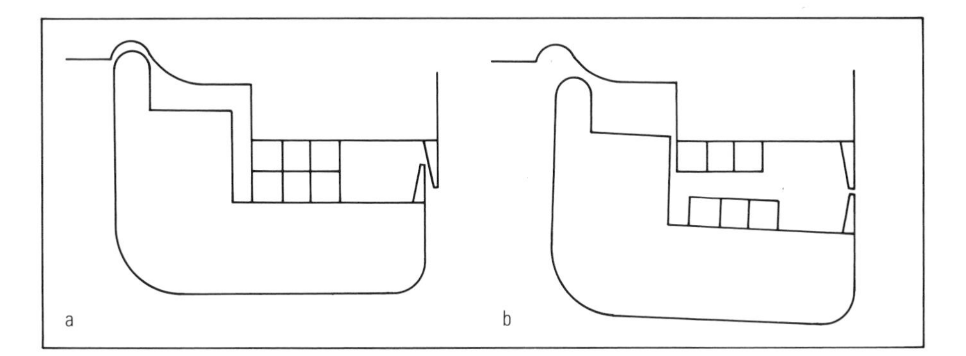

FIG. 11.3 *Diagrams of human jaws with (a) the upper and lower molars in contact and (b) the incisors meeting edge to edge. The canines and premolars have been omitted for clarity.*

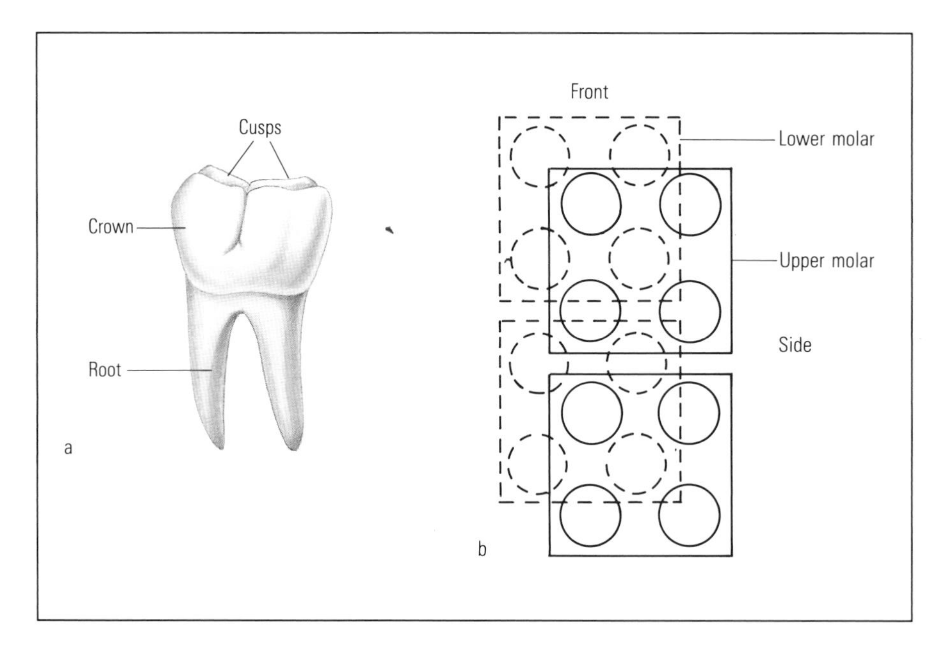

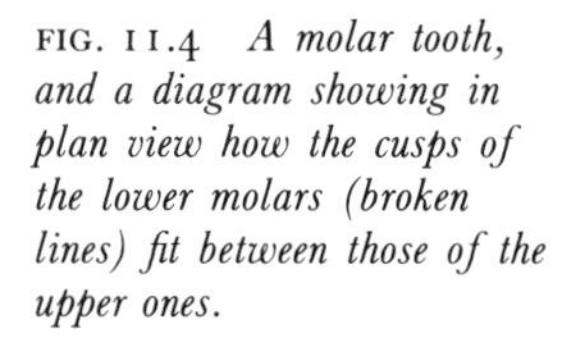
FIG. 11.4 *A molar tooth, and a diagram showing in plan view how the cusps of the lower molars (broken lines) fit between those of the upper ones.*

you close your incisors to an overlapping position or make them meet edge to edge.

Premolar and molar teeth are very different from incisors: they are squarish blocks with two (premolars) or four (molars) cusps on them. The cusps of the lower teeth fit very neatly between those of the upper ones, when the mouth is closed (FIG. 11.4b). You will realize how good the fit normally is if your dentist spoils it by finishing off a filling even slightly too high.

Chewing movements have been investigated in various ways. Australian aborigines chew with their lips parted, and the movements of their incisor teeth have been filmed. In other investigations X-ray cinematography has been used to observe the movements of the teeth inside the closed mouth. In yet others, pointers or remote sensing devices have been attached to the jaws. All these methods show that the lower jaw moves up and down and from side to side, in an elliptical path (FIG. 11.5a). Most people chew sometimes on the left side of the mouth but more often on the right, changing periodically from one to the other. The side where it is currently being chewed is called the working side, and the other the balancing side. The lower jaw moves over towards the working side as it rises in each chewing cycle and towards the balancing side as it falls. If it moved vertically up as in FIG. 11.5b it would merely crush the food. If it moved at the angle of the cusp surfaces (FIG. 11.5c), these surfaces would slide over each other, grinding the food much as corn is ground between millstones. The actual jaw movement seems to have both crushing and grinding effects.

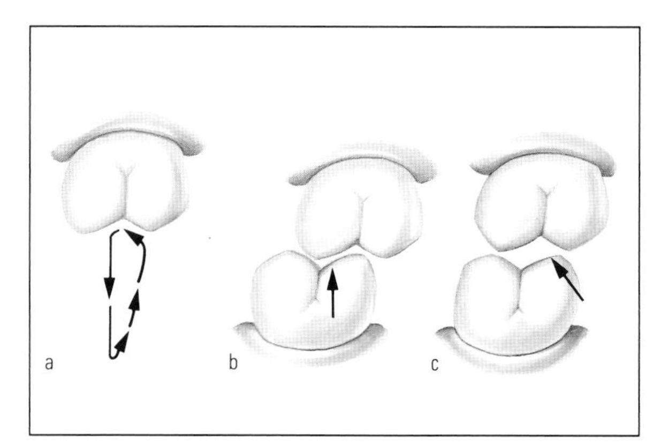

FIG. 11.5 *(a) shows the path of a lower molar relative to an upper one, in a typical chewing cycle. A directly vertical movement (b) would merely crush food but a sloping one (c) would have a grinding effect.*

Five pairs of muscles run between the skull and the lower jaw (FIG. 11.6). The temporalis, masseter and medial pterygoid are all jaw closers. (The masseter lies on the outer face of the jaw and the medial pterygoid on the inner.) The cross-sectional areas of these muscles are about 9, 11 and 7 square centimetres, respectively, so they presumably exert forces in about those proportions. The lateral pterygoid is a much weaker muscle (3 square centimetres) on the inner side of the lower jaw, which serves to pull it forward to bring the incisors edge to edge. The digastric is a mouth-opening muscle that consists of two fleshy bellies connected in series by a tendon. It passes through a loop of tendon-like tissue, connected to the small bone (the hyoid) that lies below the tongue. This prevents it from taking a straight line from skull to jaw.

We can exert bigger forces with our molars than with our incisors and canines, because they are nearer the jaw joint. Similarly, the effectiveness of nutcrackers depends on the nut being close to the pivot. In an experiment in which a man bit a force transducer he was able to exert up to 750 newtons with his second molars (this was probably about equal to his body weight), but only 450 newtons with his canines.

Although we chew on only one side of the mouth at a time, we use the jaw closing muscles of both sides. You can demonstrate this by holding a toffee between the molars of one side of the mouth. Put your fingers on the sides of your lower jaw so that you can feel the masseter muscles. When you bite hard on the toffee you will probably be able to feel both masseter muscles swelling. The same thing can be shown by recording the electrical activity of the muscles, by means of electrodes stuck to the overlying skin. However, the electrical records show that the muscles on the working side are generally more active than on the balancing side.

This may not seem surprising, and there is a good reason for it. In FIG. 11.7 the muscles are pulling upward on the jaw (labelled 'muscle force') and the food is pushing down on it ('bite force'). Forces must also act at at least one of the jaw joints. The muscle force shown in this diagram represents the combined effect of all the jaw muscles. If the muscles of the working and balancing sides were contracting equally strongly, it would act midway between the two sides. However, it is shown nearer the working side, as when the working side muscles were contracting more strongly. More precisely, it is shown just sufficiently nearer the working side to make it pass through the broken line which connects the molar where the bite force acts, to the balançing side jaw condyle. All the forces shown in the diagram intersect this line, so exert no moment about it. No

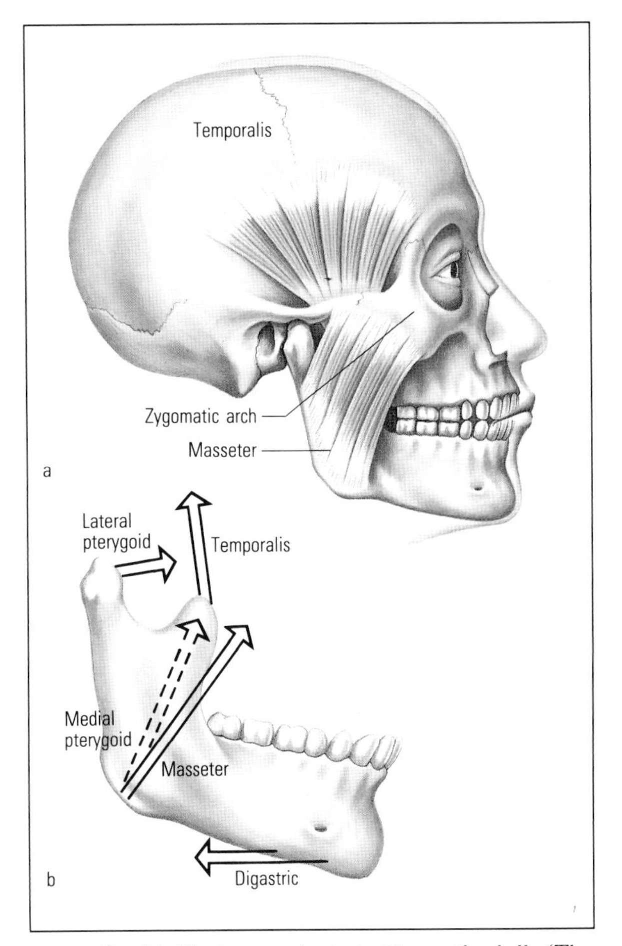

FIG. 11.6 *(a) The jaw muscles in position on the skull. (The ones that attach to the inner side of the jaw are, of course, hidden.) (b) The arrows show the directions in which the jaw muscles pull.*

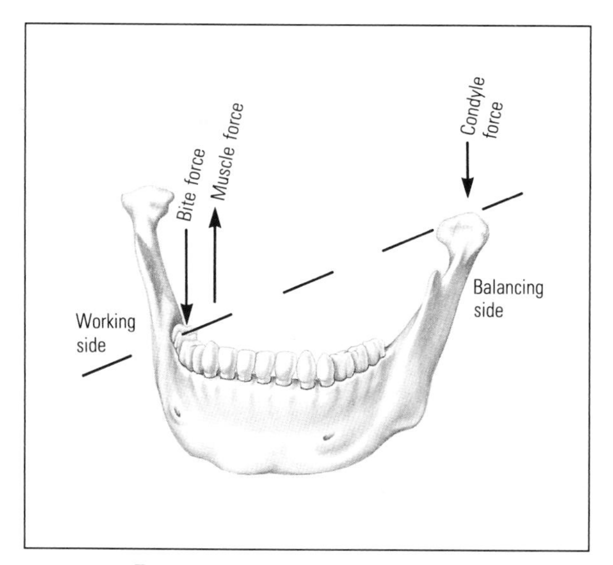

FIG. 11.7 *Forces on a lower jaw that is biting with the right molars.*

joint force is shown on the working side, because if there was one it would exert a moment about the line, which could not be balanced by any of the other forces.

If the balancing side jaw muscles contracted less strongly, shifting the muscle force in FIG. 11.7 even further towards the working side, it would exert a moment about the line which would need a downward force on the working side condyle to balance it. Both jaw condyles would be pressed tightly into their sockets. On the other hand, if the balancing-side muscles contracted more strongly, the muscle force would act on the balancing side of the line and would exert a moment about it that could only be balanced by an *upward* force on the working side joint. The condyle of this side would not be pushed into its socket: rather, it would be pulled out of its socket so that its ligaments were in tension. If we used the jaw muscles of both sides equally while chewing on one side of the mouth, we might be in danger of dislocating our jaws.

FIG. 11.7 shows no force at the working side joint. It is probably more usual to have some downward force there, though less than on the balancing side. The smaller force on the working side seems to explain an observation that might otherwise seem puzzling. Patients who have a fracture on one side of the jaw, near the condyle, prefer to chew on the damaged side.

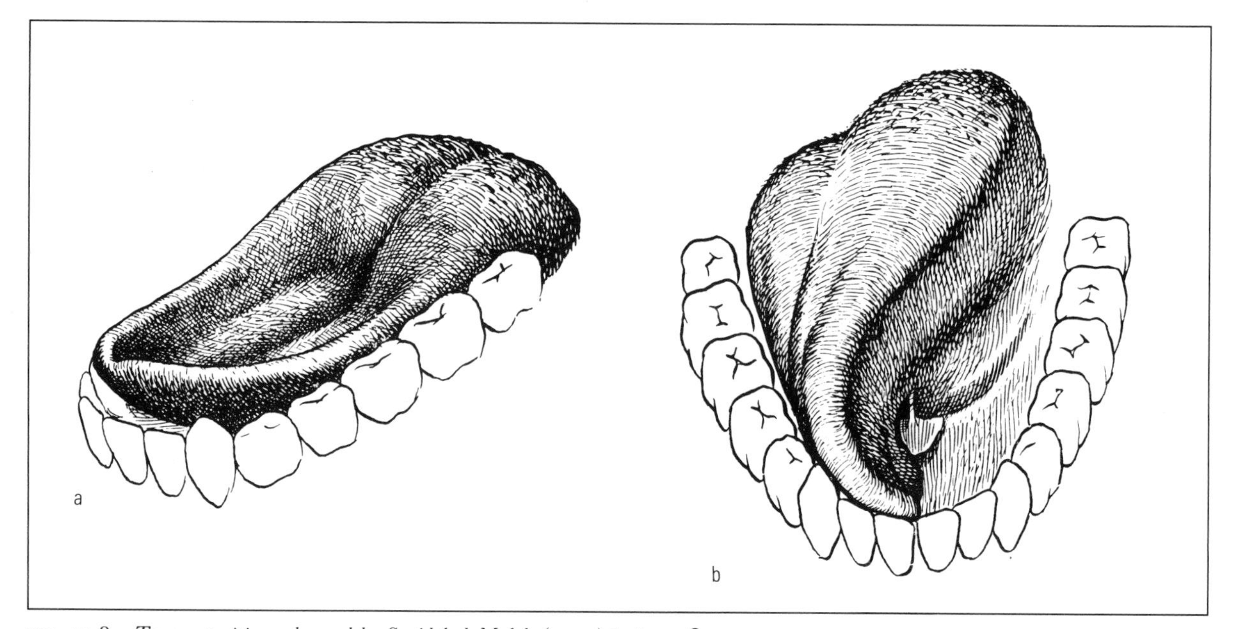

FIG. 11.8 *Tongue positions observed by S. Abd-el-Malek (1955)* J. Anat. **89**, 250–254.

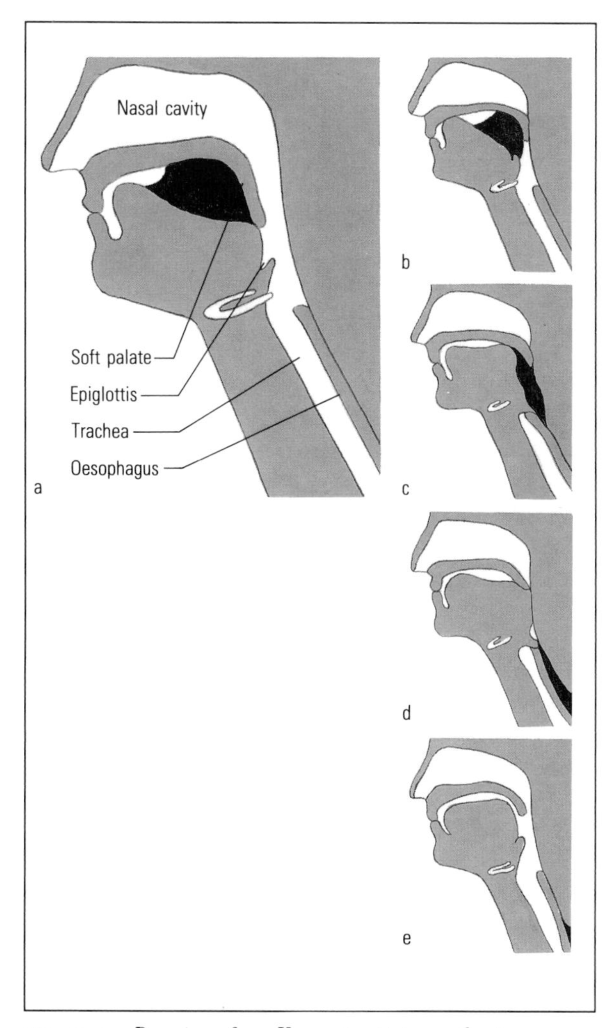

FIG. 11.9 *Drawings from X-ray cine pictures of a person swallowing food mixed with barium sulphate. The interval between successive pictures is 0.07 seconds. After H.W. Davenport (1982)* Physiology of the Digestive Tract *ed. 5. Year Book Medical Publishers, Chicago.*

Food is placed between teeth that are to chew it, by movements of the tongue and cheeks. FIG. 11.8 shows tongue positions in people who had teeth missing: the investigator parted their lips with forceps and peered through the gaps between the teeth. In FIG. 11.8a the tongue has formed a trough to gather food together, prior to moving it. In (b) it has pushed the food over to the right and is keeping it there for chewing.

These and other tongue movements are possible because of the way in which the muscle fibres of the tongue are arranged. If you eat ox tongue, compare a slice with a slice of beef or ham. Large areas of the beef or ham slice have all their muscle fibres running parallel to each other (these areas are cuts through individual muscles or parts of muscles), but everywhere in the tongue there are interwoven fibres running in various directions. Both human and ox tongues have longitudinal fibres running along them from base to tip, transverse fibres running across them and vertical ones running from top to bottom. If the longitudinal fibres shorten the tongue must get wider or thicker or both, because its volume cannot change. If the transverse fibres shorten it must get longer or thicker. Contraction of the longitudinal fibres of its left side only will bend it to the left and contraction of the fibres near its upper surface will make it form a hollow. The base of the tongue can also be moved, by muscles that run between it and the lower jaw and skull.

The process of swallowing has been studied by X-ray cinematography. The food was mixed with barium sulphate (which is opaque to X-rays) to make it as well as the skeleton visible. FIG. 11.9 was obtained in this way.

There are three possible routes for food or drink, out of the back of the mouth: it may move up into the nasal cavity or down either into the trachea (windpipe) or into the oesophagus (gullet) (FIG. 11.9a). To get it to the stomach as required it must be sent down the oesophagus and kept out of the other two routes. FIG. 11.9 shows how this is done. The palate separates the nasal cavity from the mouth. Its front part is stiffened by bone but the rear part (the soft palate) is flexible and muscular. Initially (FIG. 11.9a) the soft palate rests against the back of the tongue, closing the back of the mouth cavity. It is raised by tightening its muscles, and probably also by the food pressing against it, as the tongue pushes the food back for swallowing (b). This closes the opening to the nasal cavity. As the food moves back it presses on the epiglottis, a flexible flap of tissue at the entrance to the trachea, which bends over to close this off (c). There is now only one path open to the food when the tongue rises to force it right out of the mouth: it must go down the oesophagus (d). After it has passed, the epiglottis springs up again by elastic recoil. A wave of muscular contraction travels down the oesophagus, pushing the food in front of it (e). This wave may not be necessary in normal drinking, when gravity makes the liquid flow down the oesophagus, but it is possible to drink while standing on one's head.

Competitive beer drinkers may seem simply to pour beer down their throats, but do not depend entirely on gravity. X-ray cine pictures of an 'expert' (drinking barium sulphate suspension instead of beer) showed his tongue moving much as in FIG. 11.9, making about one cycle of movements each second.

Food is moved by muscular waves in the stomach and intestines, as well as in the oesophagus. The process, called peristalsis, is illustrated in FIG. 11.10. The muscle fibres concerned run circumferentially around the organ, so they constrict it when they shorten. These fibres are described as 'smooth' to distinguish them from the 'striated' fibres of skeletal muscles. Unlike striated fibres (FIG. 2.12) they do not appear striped under the microscope because their fibres are arranged in a much less orderly way. Also, they are not under voluntary control. You cannot choose to send a wave of contraction along your intestine in the way that you can choose to bend a finger. There are similar muscle fibres in the walls of blood vessels.

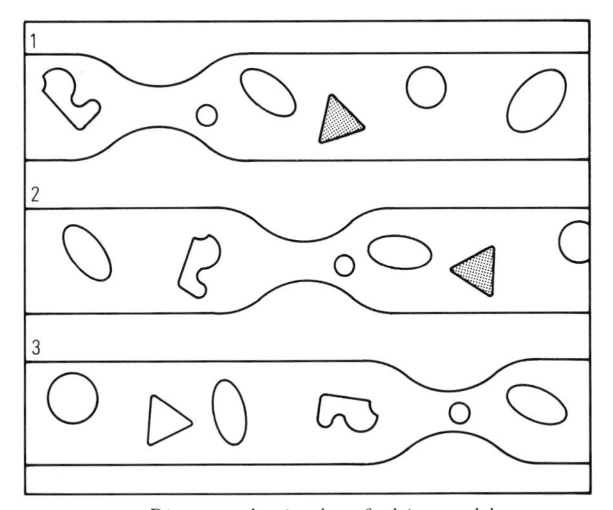

FIG. 11.10 *Diagrams showing how food is moved by peristalsis.*

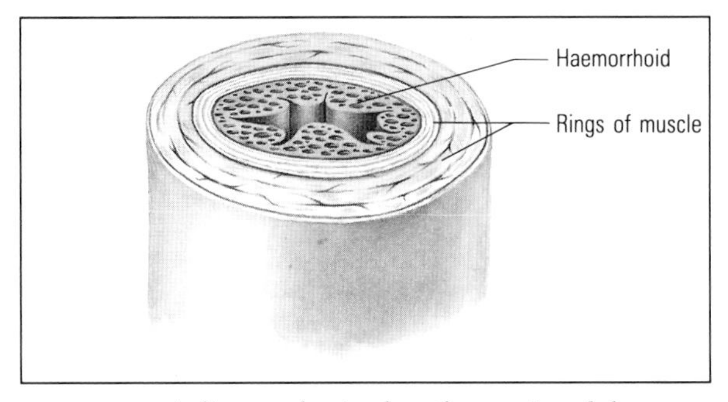

FIG. 11.11 *A diagram showing how the anus is sealed.*

Defaecation does not depend entirely on peristalsis, but uses the muscles of the body wall to build up pressure in the body cavity and force the faeces out.

The anus needs a watertight seal to prevent leakage. It is surrounded by two rings of muscle (FIG. 11.11), an inner ring of smooth muscle and an outer one of striated muscle, which contract to close it. Inside the inner ring are the haemorrhoids, spongy pads of tissue swollen with blood. They sometimes give trouble and used to be regarded as pathological but are now recognised as a normal part of the body with an important function. The blood comes to them through large vessels so the pressure inside them is close to the pressure in the arteries, and they seal the anus tightly.

This last paragraph is about an organ which has nothing to do with the main subject of this chapter but is included here for convenience of comparison with the haemorrhoids. Like them, the penis consists of spongy, blood-filled tissue, but it is filled with blood at high pressure only when erected. Its tissue includes a large proportion of collagen fibres which are slack when it is flaccid, but taut when it is erected. It becomes rigid when erected because some of the fibres would then have to be stretched, to bend it. This is the principle of inflatable structures generally. Rubber dinghies and the escape chutes of aircraft become rigid when inflated with air. Inflatable structures are sometimes used as roofs for temporary buildings. Even herbaceous plants are inflatable structures: their cells are normally filled with sap under pressure, and they wilt when sap pressure falls.

CHAPTER TWELVE

BREATHING

We have two cavities in the trunk, the thoracic cavity which contains the heart and lungs, and the abdominal cavity which contains stomach, intestines, liver and other guts (FIGS. 12.1 and 12.3). They are separated by a partition, the diaphragm. The oesophagus and major blood vessels pass through this so that food can travel from mouth to stomach and blood can circulate round the body, but the diaphragm is tightly sealed around these tubes, so that fluid in the cavities themselves cannot leak from one to the other. The outer wall of the thorax is stiffened by ribs which are jointed to the thoracic

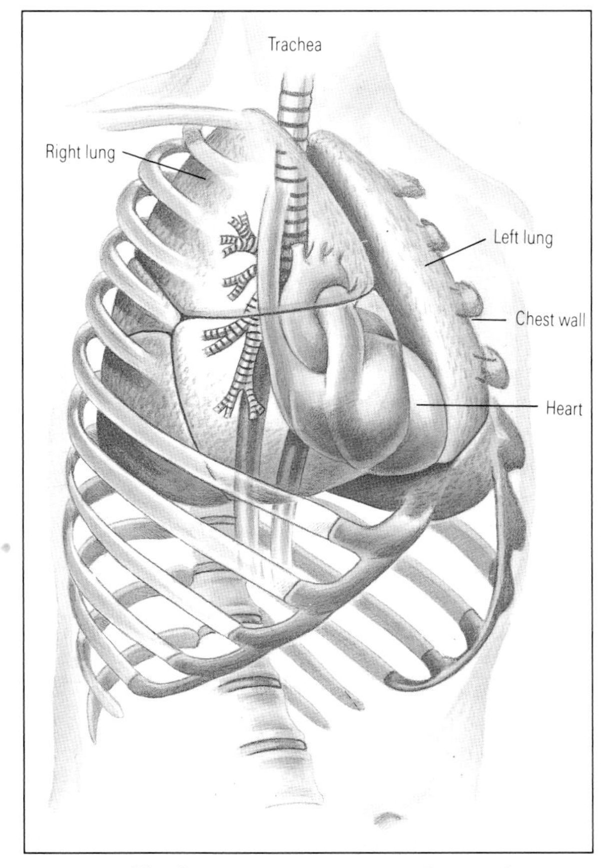

FIG. 12.1 *The thorax made transparent to show the heart and lungs.*

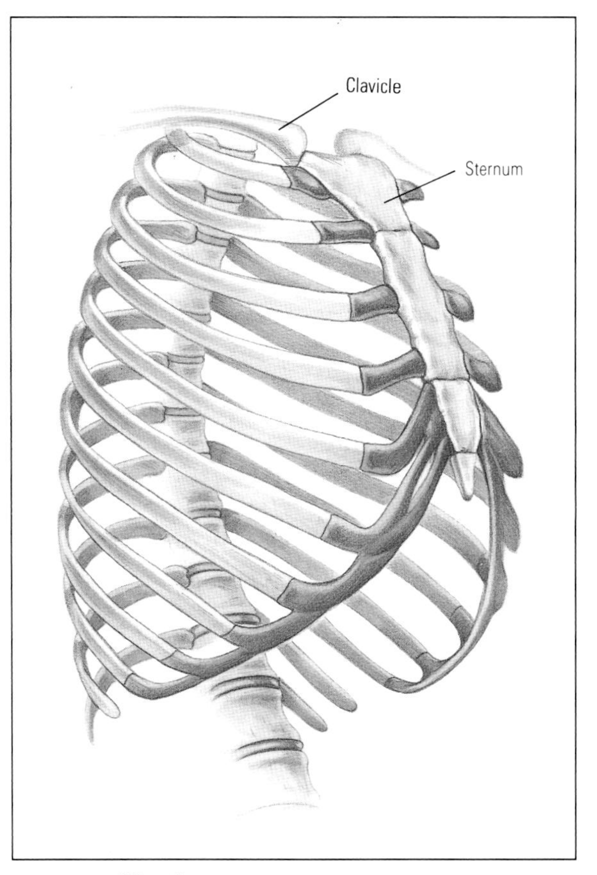

FIG. 12.2 *The rib cage.*

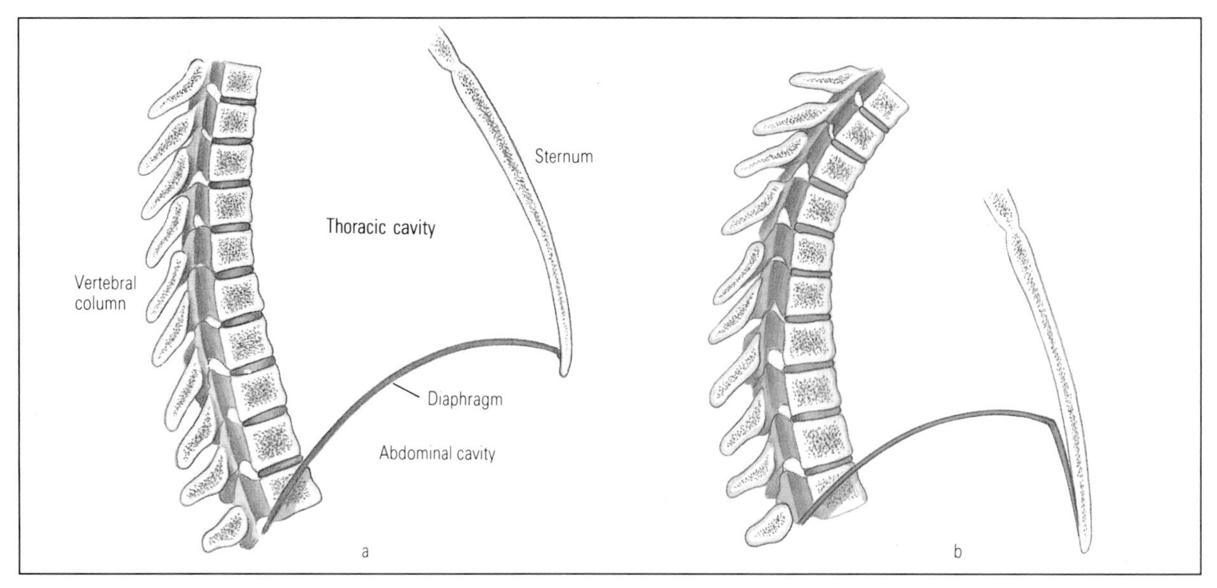

FIG. 12.3 *Tracings of X-ray pictures of the chest from the side with the lungs fully expanded (a) and after breathing out (b). After N.M.T. Braun, N.S. Arora & D.F. Rochester (1982)* J. appl. Physiol. **53**, *405–412*.

vertebrae behind and to the sternum (breastbone) in front. FIG. 12.2 shows the arrangement. Notice that part of each rib, near the sternum, is made of cartilage instead of bone, and that the last few ribs do not reach the sternum, or connect with it only indirectly by forming joints with the ribs above them. The diaphragm is attached to the lower edge of the rib cage and to the first few lumbar vertebrae.

Bones show up in radiographs because they are more opaque than flesh to X-rays, and air-filled spaces also show up because they are more transparent: bones appear white in the radiograph, flesh grey and the lungs blacker. The position of the diaphragm can be found because the lungs rest directly on top of it. FIG. 12.3a was traced from X-ray pictures from the side of someone who had breathed deeply in, inflating the lungs as much as possible. It shows the rib cage expanded and the diaphragm forming a rather flat dome. FIG. 12.3b shows what happens on breathing out: the rib cage is narrower both from front to back and from side to side, and the diaphragm reaches much further up into it. There is much less space for the lungs to occupy in (b) than in (a): the volume of air in them was estimated from measurements on the pictures to be only about 62% as much as in (a).

The centre of the diaphragm is a sheet of collagen fibres but there are muscle fascicles running radially all around the edge. This is striated muscle like the muscles that work our skeletons, not unstriated like the muscle fibres in the walls of intestines and blood vessels. The fascicles must be 70% longer in FIG. 12.3(b) than in (a), to allow the diaphragm to move so far up. Notice that when they are as long as this, the edges of the diaphragm lie flat against the inner surfaces of the last few ribs. Electrodes pushed through the body wall into the diaphragm have been used to record its electrical activity. The records show that the diaphragm muscles contract (enlarging the thoracic cavity) when we breathe in and relax (or at least are less active) when we breathe out.

When we breathe in and the diaphragm moves down, it pushes the guts in the abdominal cavity downwards. The wall of the abdomen is flexible enough to accommodate that: breathing in makes the belly swell a little.

The movements that enlarge our rib cages, when we breathe in, can be thought of as a combination of

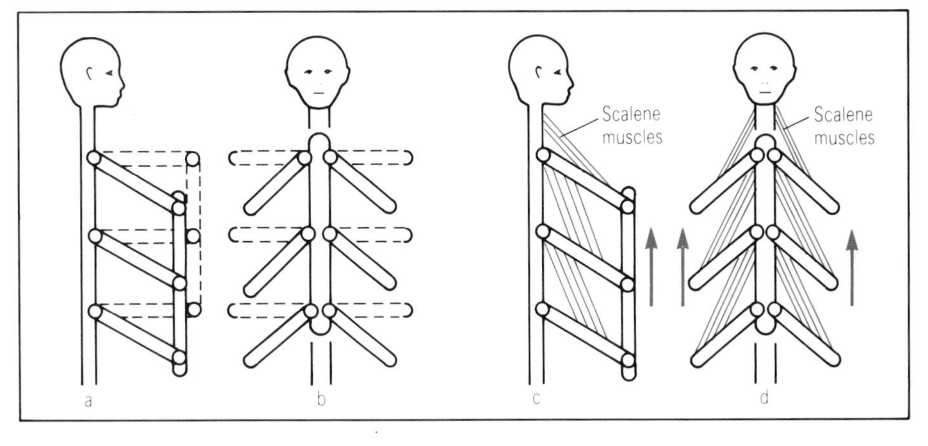

FIG. 12.4 *Simplified diagrams of rib movements.*

two simpler movements. Imagine ribs that were attached to vertebrae and sternum by hinge joints with axes that ran transversely across the body (FIG. 12.4a). The volume enclosed by the ribs would be greater when they were horizontal, than when they were sloping. Now imagine ribs that were attached to vertebrae and sternum by hinges with horizontal front-to-rear axes. They could rise and fall as in FIG. 12.4b and the volume enclosed would again be greater when they were horizontal, than when they sloped.

Raising the ribs as in FIG. 12.4a enlarges the thorax if the front end of each rib (at the sternum) is initially lower than the back end (at the backbone). Raising them as in FIG. 12.4b will enlarge it if the middle part of each rib (at the side of the body) is initially lower than the ends. Our ribs are actually curved in a complex way so that the front end of each is lower than the rear, and the side is lower than either (FIG. 12.2), so both movements will be effective. Their actual movements are a combination of the two shown in the simple diagram because the hinge joints between ribs and vertebrae have neither transverse axes (as in FIG. 12.4a) nor front-to-rear ones (as in FIG. 12.4b), but are inclined at an angle between the two (as shown in FIG. 12.5). When we raise our ribs to breathe in the thorax gets wider both from side to side and, as FIG. 12.3 shows, from front to back.

The hinge joint between each rib and the vertebrae actually involves two cartilage-covered areas of contact, one at the extreme end of the rib where it presses against the body of a thoracic vertebra (or against the bodies of two adjacent vertebrae), and one a little back from the end where it touches a transverse process on the neural arch (FIG. 12.5). The hinge axis runs parallel to a line joining these points of contact.

Look at the highly simplified diagrams (FIG. 12.4). Diagram (c) shows muscle fascicles running diagonally between adjacent ribs: if they shorten, they will raise the ribs as in (a). The scalene muscles, running from the neck vertebrae to the first rib, will help them. Diagram (d) shows another set of muscle fascicles between adjacent ribs and between them

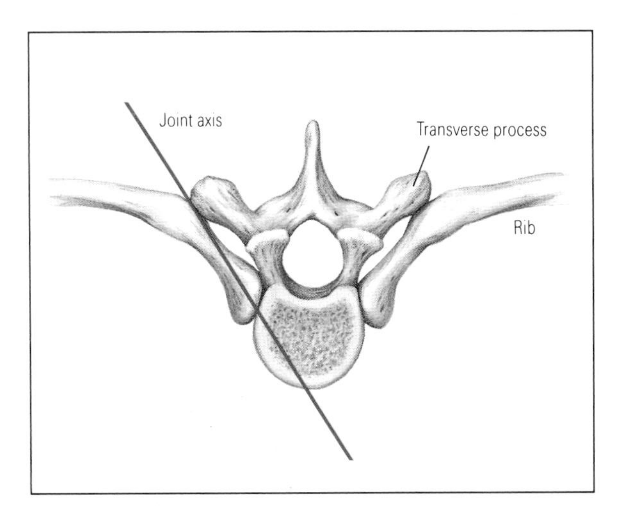

FIG. 12.5 *The hinge joint between a rib and the vertebrae.*

and the sternum: if they shorten while the scalene muscles prevent the sternum from moving downwards, they will raise the ribs as in (b).

There are two layers of muscle connecting each rib to the next. The external intercostal muscles run as shown in FIG. 12.4c while the internal intercostals run in the other diagonal direction and have the opposite effect on the ribs. The intercostal muscles shown in FIG. 12.4d are part of the internal layer, and have an external layer (with the opposite effect) over them. There are abdominal muscles that can pull the sternum downwards as well as scalene muscles that pull it up. Between them, these muscles cause the rib movements of breathing. Records of the electrical activity in them show (as you would expect) that the muscles that expand the rib cage and those with the opposite effect are active alternately. However, elastic recoil also plays a part in emptying the lungs. Just as a rubber balloon empties itself if the air is allowed to escape, so also (much less forcibly) do the lungs.

Adult men can inflate their lungs to a total volume of about seven litres, by breathing in as deeply as possible. They can reduce their volume to about two litres by breathing out forcibly. The difference (about five litres) is the volume of the biggest possible breath, but we seldom take breaths as big as that. Measurements made in various ways, for example by having the subject breathe out through a mouthpiece with a built-in flowmeter, show that even in heavy exercise, the volume of air moved in each breath is seldom more than about 3.3 litres.

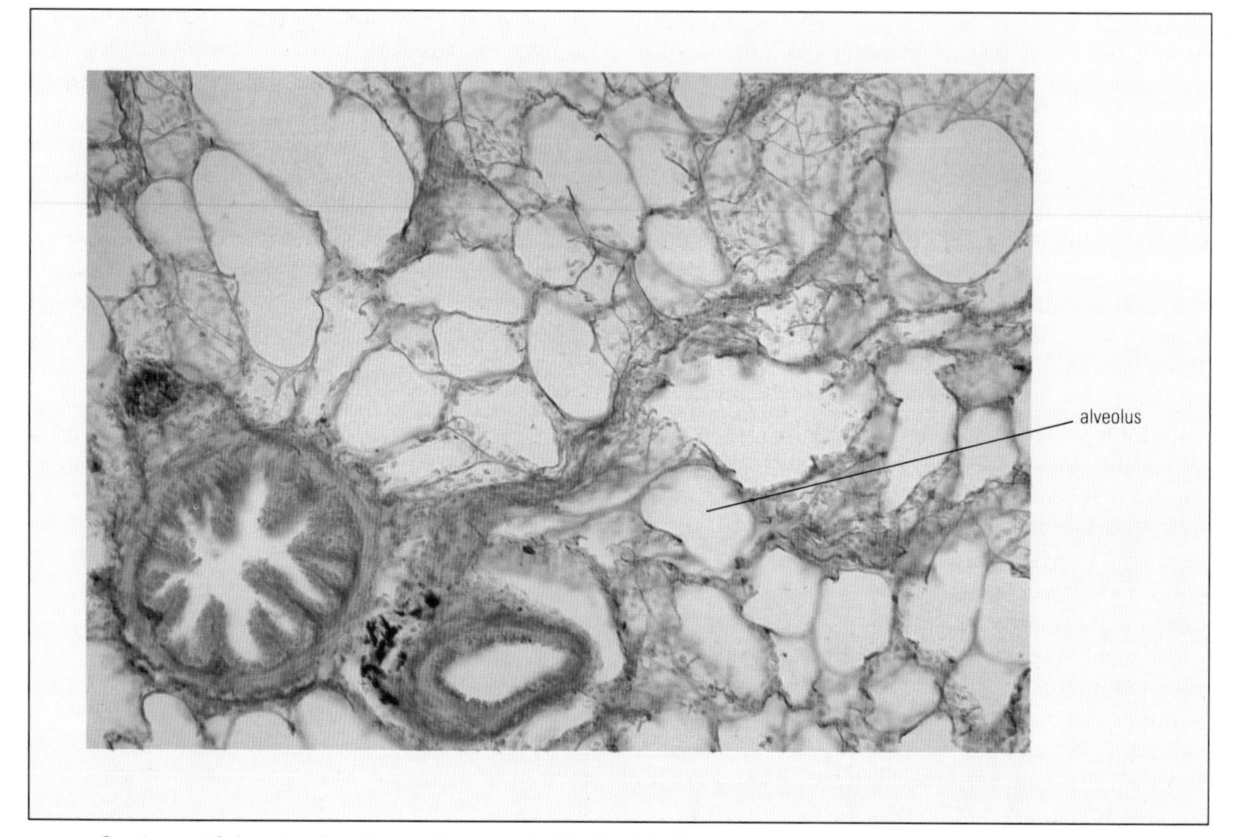

FIG. 12.6 *A magnified section through a cat lung, supplied by Dr P.J. Evennett, Department of Pure and Applied Biology, University of Leeds.*

Healthy young men take 10–20 small breaths per minute when resting, and up to about 40 large (three litre) breaths in heavy exercise. Thus they pass up to about 120 litres of air per minute through their lungs. That volume of fresh air contains about 25 litres of oxygen, but the lungs can extract no more than about four litres of oxygen per minute from the air. (A bigger *proportion* of the oxygen can be extracted from the smaller amounts of air that are breathed at rest.)

The trachea or windpipe (the stiff-walled tube that you can feel at the front of your throat) carries air from the back of your mouth down into the thorax. There the trachea divides into two bronchi (one for each lung) which divide into many smaller branches that go to the various lobes of the lungs (FIG. 12.1). All these tubes have incomplete rings of cartilage embedded in their walls, so they are flexible but cannot be squashed flat, like the hoses of vacuum cleaners.

Further divisions of the airways end eventually in tiny pockets called alveoli (FIG. 12.6): their diameters, in three-quarter filled human lungs, average about 0.3 millimetres. In this way, an enormous surface area is packed into the limited volume of our lungs. A 0.3 millimetre sphere has a volume of 0.014 cubic millimetres and a surface area of 0.28 square millimetres. The three-quarter filled lungs would contain about 4.8 litres of air but some of this would be in the bronchi and their branches, and only about 3.2 litres (3.2 million cubic millimetres) in the alveoli. That means that there must be about 3 200 000/0.014 = 230 million alveoli, with a total surface area of 64 million square millimetres = 64 square metres. That was a very rough calculation: a slightly more elaborate and (one hopes) more accurate one gave 81 square metres. For comparison, the area of a tennis court is 261 square metres, equivalent to only 3.2 pairs of lungs. A huge area is needed to allow oxygen to diffuse rapidly enough into the blood and carbon dioxide to diffuse out.

A huge area would not by itself be enough to make fast diffusion possible. The air must be brought very close to the blood so that the barrier of tissue, through which the gases must diffuse, is very thin. The walls of the alveoli are extremely delicate, and permeated by a network of tiny blood vessels, so that over much of the area of the alveoli the air comes within one fifth of a micrometre (one five thousandth part of a millimetre) of the blood.

The small size of the alveoli leads to a difficulty. The surfaces are presumably moist, but surfaces between air and water behave as if they were in tension, and the tension will tend to collapse the alveoli. To understand the consequences, we need to know something about surface tension. This is the force that makes soap bubbles spherical and enables some insects to walk on water.

Tension in a stretched rubber membrane (for example, the wall of an inflated balloon) tends to reduce the membrane's area. Similarly, surface tension tends to reduce the surface areas of bubbles. It makes them spherical because a sphere has a smaller surface area than any other shape of the same volume. The surface tension at clean surfaces between water and air is 0.07 newtons per metre: that means that if you drew a one-metre line on the surface of water, the tension acting across it would be 0.07 newtons.

The tension in the surface of a bubble tends to make the bubble smaller, increasing the pressure of the gas inside. Imagine a bubble cut in half (FIG. 12.7). The two halves are held together by surface tension and pushed apart by the pressure inside them. The surface tension force is proportional to the circumference of the bubble, and so to its diameter, but the pressure force is proportional to

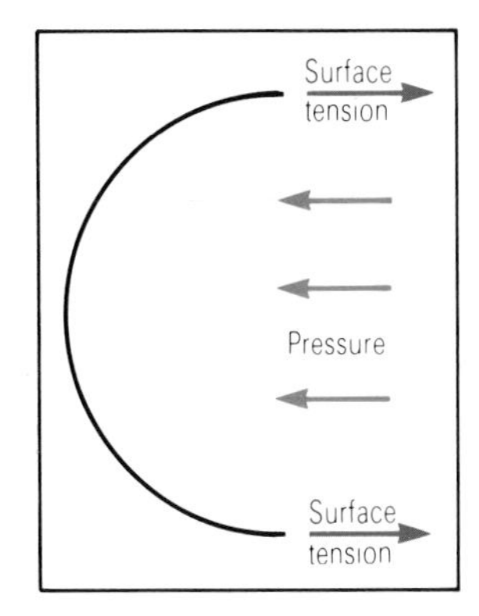

FIG. 12.7 *The forces on half an air bubble.*

the area of the section through it, and so to diameter squared; and the two forces must balance. That means that the smaller the bubble, the bigger the pressure difference must be between inside and outside. The equation, for a gas bubble submerged in liquid, is

pressure difference = 4 × surface tension/diameter

Think of an air bubble the size of an alveolus (0.3 millimetres diameter) in water of surface tension 0.07 newtons per metre. The pressure difference would be 4 × 0.07/0.0003 = 900 newtons per square metre, equal to the hydrostatic pressure at the bottom of a 9 centimetre column of water. That raises no problem so far: it is less, as we will soon see, than the pressures needed to inflate lungs. Remember, though, that we did the calculation for a three-quarters inflated lung. Breathing out would make the alveoli smaller and the pressures due to surface tension larger. If any of them ever got too small, the pressures needed to keep them inflated might become impossibly large: they would collapse completely and be very hard to reinflate. Similarly, (but for a rather different reason) blowing up a rubber balloon is very difficult to start with but gets easier as it inflates.

FIG. 12.8 shows the results of an experiment with lungs dissected out of a cat. First, the lungs were inflated with air and allowed to deflate, while the pressure required was measured. Next they were emptied of air by using a vacuum pump and refilled with a weak salt solution. (The concentration of salt was chosen to match the concentration in the blood.) Less than half as much pressure was needed to inflate the lungs with this solution than to inflate them to the same volume with air.

An important difference between filling with air and with salt solution, is that in the second case there is no fluid-gas surface: the effect of surface tension has been eliminated. The large difference between the pressures needed to fill the lungs with air and with salt solution is presumably due to surface tension.

There is another difference between the pressure records. The record for filling with salt solution

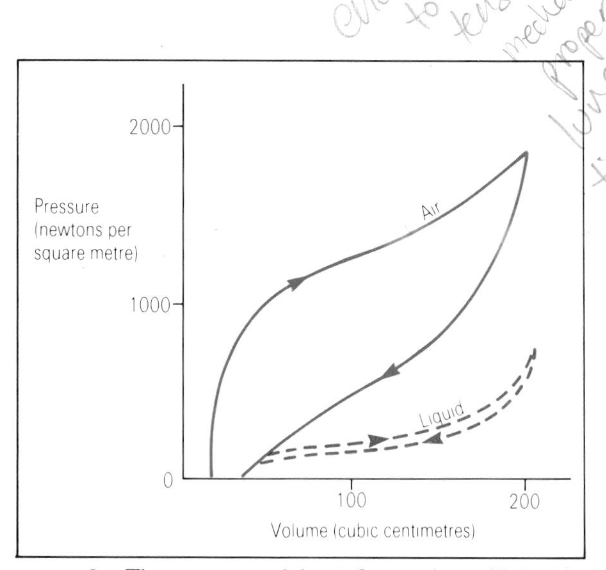

FIG. 12.8 *The pressures needed to inflate cat lungs filled with air and with liquid, based on a graph by E.P. Radford (1957) in J.W. Remington (ed.)* Tissue Elasticity. *American Physiological Society, Washington.*

forms a narrow loop showing that most of the energy used in stretching the lungs is returned in their elastic recoil. (The graph shows pressure plotted against volume but the principle is the same as for graphs of force against extension for stretched tendons, FIG. 3.5). The record for filling with air, however, makes a very large loop, showing that much of the energy is lost. This happens only when the lung is inflated and deflated (as in this experiment) through a very large range of volumes. Smaller volume changes like those of normal breathing involve *much* less energy loss. However, the big losses in the experiment show that there is something odd about the surface tension in the lungs.

Fluid squeezed out of lungs forms a froth of very stable bubbles, which suggests that it contains material with soap-like properties, a surfactant. When the foam is dried, a waxy substance remains. It is dipalmitoyl phosphatidylcholine (DPPC for short). Like other surfactants it accumulates at an air-water surface, forming a continuous layer just one molecule thick. This reduces the surface tension.

FIG. 12.9 shows one of the ways in which the surface tension of a surfactant film can be measured.

The float separates a clean water surface from one covered by the film so it is pulled one way by the surface tension of water and the other by that of the film. The force needed to hold it stationary (which is measured) represents the difference between the two surface tensions. A barrier on the other side of the film can be pushed towards the float to reduce the film area or away from it to increase the area.

When a DPPC film is tested with the barrier stationary, it reaches equilibrium at a surface tension of 0.025 newtons per metre, about one third of the surface tension of pure water. The surface tension in our lungs must remain close to that when we are breathing gently, so the pressure due to the surface tension must be only about one third of our initial estimate. When the barrier in the apparatus is moved back and forth, greatly increasing and then decreasing the area of the DPPC film, its surface tension fluctuates as shown in the graph (FIG. 12.9). It approaches 0.050 newtons per metre as the area increases and falls close to zero as it decreases. This is because molecules take some time to move in and out of the surface film. Increasing the area makes gaps appear in the film, so the surface tension approaches that of water. Decreasing it packs the molecules of the film more tightly together so that further decrease of area is resisted.

When lungs are inflated, the surfactant-covered area increases and when they are deflated it decreases. The large loop in the surface tension-area graph for surfactant film (FIG. 12.9) seems to explain the large loop in the pressure-volume graph for air-filled lungs (FIG. 12.8). However, there is some controversy about this because the sequence of enlargement of the alveoli as the lung inflates is not exactly reversed as it deflates.

Infants are born with their lungs deflated, containing only fluid. They must be inflated with air within a few minutes so that the child can start breathing. If there were no surfactant, very large pressures would be needed to start the inflation. If some alveoli got larger than the rest, the pressure needed to inflate them further would fall: smaller (high pressure) alveoli would collapse, driving the air into larger (low pressure) ones. Parts of the lung would inflate while others remained collapsed and would be very difficult to inflate. This actually happens in babies suffering from respiratory distress syndrome, who have difficulty in breathing and often die if not suitably treated. Post mortem examination shows that some parts of the lungs are still filled with fluid while others are excessively distended. Further tests show that the surfactant is deficient: when films of it are compressed, their surface tension remains above 0.02 newtons per metre, failing to fall to the very low values shown in FIG. 12.9. Babies suffering from the syndrome have been treated successfully by putting synthetic surfactant into their lungs. The lungs of healthy babies inflate uniformly because if some alveoli start inflating faster than others the surface tension in them rises, resisting further inflation.

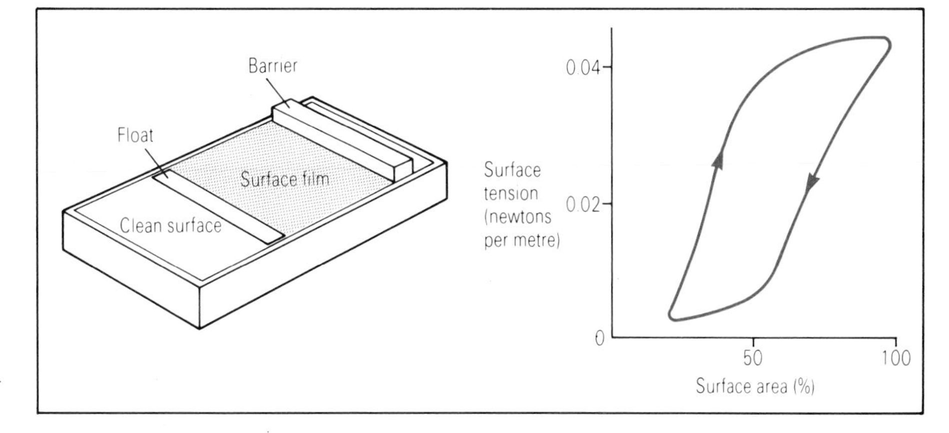

FIG. 12.9 *Apparatus for measuring the surface tension of a surfactant film, and a graph of surface tension against area for a film of lung extract. The graph is after J.A. Clements (1962)* Scient. Am. **207**, *(6), 121–130.*

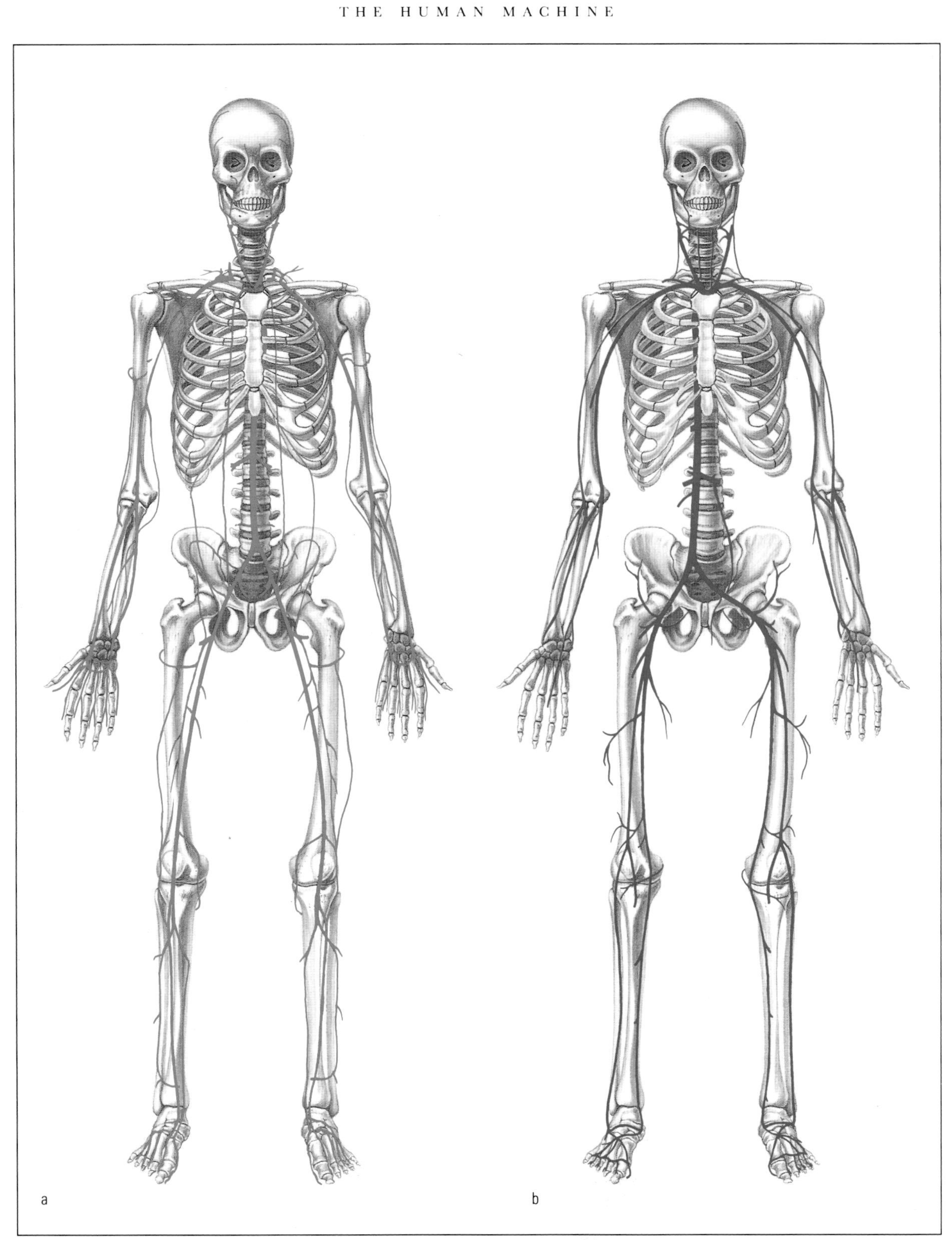

FIG. 13.1 *The principal arteries (a) and veins (b) of the body, excluding those of the gut and liver.*

CHAPTER THIRTEEN

BLOOD CIRCULATION

Blood collects oxygen from the lungs and delivers it all around the body to the tissues where it is needed for metabolism. It carries the carbon dioxide that is produced by metabolism back to the lungs, where it diffuses out into the air. It carries the products of digestion from the gut to other parts of the body, and waste products to the kidneys. It also carries the hormones that influence our growth, metabolism and behaviour. It does not of course offer a direct messenger service: if glucose from the liver (for example) is needed in the leg muscles it is not carried there directly but circulates to all parts of the body. Once in the blood a particular molecule must arrive eventually (possibly after many circuits through the heart) at the organ where it is needed.

This flow is driven by the heart, whose walls consist of muscle of a special kind that will continue beating without needing a stimulus from its nerves to start every contraction. Each contraction starts in a group of specialized muscle cells (the 'pacemaker') and spreads to the rest of the heart. Muscle fibres run in different directions in successive layers of the heart wall; some lengthwise, some circumferentially and some (corkscrew-like) in helices. The heart can be distinguished in X-ray pictures as a relatively opaque structure between the more transparent lungs, so its size can be measured even in intact living people and its volume estimated. It is much larger in endurance athletes than in sedentary people. For example, the silver medallist in the women's 400 metres freestyle swimming in the 1960 Olympic Games had a 0.87 litre heart, whereas the average for women of her size is 0.55 litres. Men have larger hearts, up to an exceptional 1.70 litres in a professional cyclist. (This fell to 0.98 litres, which is still high, after he ceased training.) Long distance runners, as well as swimmers and cyclists, have large hearts.

The heart has four chambers, two atria and two ventricles. The left atrium and ventricle are on the right in FIG. 13.2, and vice versa, because the diagram shows the heart as it would appear if the chest were cut open from in front. In each heart beat the atria contract, moving blood from them into the ventricles which in turn contract, driving it out into the arteries. The small pressure of the blood in the veins is enough to stretch the relatively thin walls of the atria and refill them, when their muscle relaxes. Contraction of the atria produces the rather larger pressures needed to refill the ventricles, which produce the very much larger pressures needed to drive blood round the body.

Valves are needed, just as they are in water pumps (FIG. 13.3), to prevent the blood from flowing the wrong way. There are valves between the atria and the ventricles that open when the atria contract but flap shut when the ventricles contract. Similar valves guard the openings from the ventricles to the arteries. All these valves work passively like those of the water pump (FIG. 13.3) but are not single flaps: each consists of two or three flaps that meet, when the valve closes, in the centre of the opening. The ones between the atria and the ventricles are prevented from turning inside out by tendons that connect them to the ventricle wall, but the valves leading out of the ventricles work well without any such protection.

Heart valves are often damaged or even destroyed by rheumatic fever but because they work passively, with no need for muscles to move them, they can be replaced by simple mechanical devices or by preserved valves from dead pigs.

FIG. 13.2 *A diagram of a human heart cut open to show the chambers and valves.*

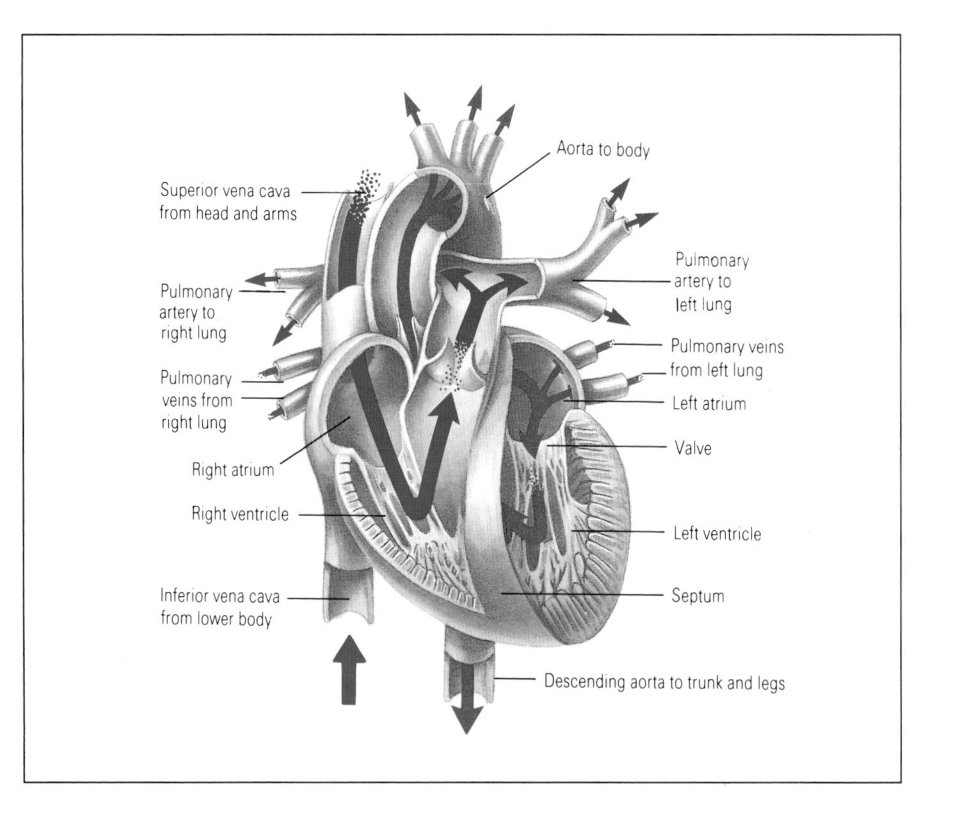

FIG. 13.3 *A water pump. The valves are simple hinged flaps that open freely to allow water to flow from left to right but close if it starts flowing the other way.*

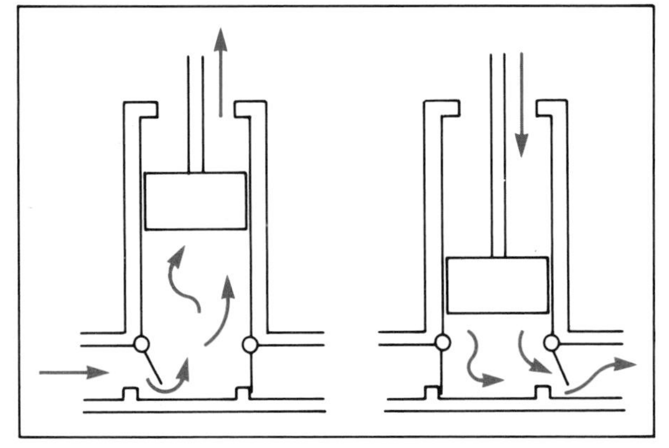

The blood flows away from the heart in arteries which have relatively thick walls (they have to withstand quite high pressures) and return to it in veins, which have thin ones. The arteries divide into successively finer branches and eventually into blood capillaries which are only about five micrometres (0.005 millimetres) in diameter, which permeate all the tissues. The downstream ends of the capillaries coalesce into small veins which in turn coalesce into successively larger ones and eventually into the very large veins which return the blood to the atria. There are valves in the veins (simple flaps like the valves of the heart) that allow flow only towards the heart. Blood from the right ventricle goes to the lungs and is returned to the left atrium. From there it goes through the left ventricle to other parts of the body and back to the right atrium. Thus the circulation can be thought of as a figure of eight with the heart at the crossing point (FIG. 13.4). On its excursions from the heart the blood goes alternately to the lungs (where it collects oxygen) and to other parts of the body (where the oxygen is used).

William Harvey, a seventeenth-century physician, is widely regarded as the first experimental scientist. His great discovery (published in 1628) was that the blood circulates, travelling always in the same direction in each blood vessel, rather than ebbing and

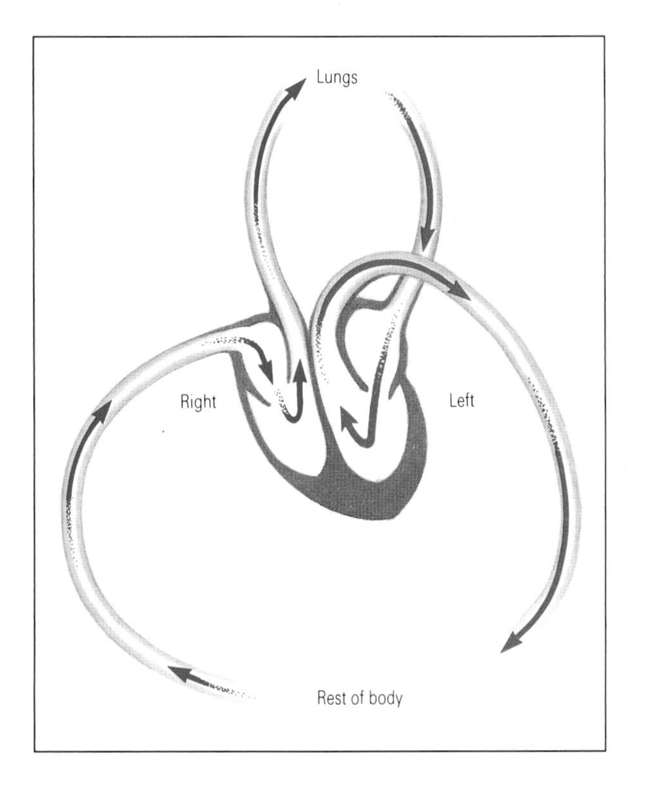

FIG. 13.4 *A simple diagram of the blood circulation. The real thing has immensely more branches.*

flowing like the tides or being produced in one part of the body and consumed in another. He described experiments showing that the arteries carry blood into the arm and veins carry it out again. He tied a ligature round an assistant's arm, preferably a lean person who had been exercising. If he tied the ligament so tight that no blood could pass, the artery above the ligament swelled. If he tied it a little less tightly, however, ('Figure 1' in FIG. 13.5) arterial blood could apparently get into the lower part of the arm but venous blood could not escape and the hand and the veins *below* the ligature became swollen. The swellings that then appeared in the veins (B, C and D in his diagram) were the sites of the valves. If Dr Harvey used his fingers to squeeze the blood from a section of a vein between two valves (OH), leaving a finger pressing on the lower end, the empty section remained empty: the upper valve (O) prevented blood from flowing backwards into it. When he raised the finger (from H) however, the empty section refilled. Blood can flow past the valves in the veins towards the heart (from H to O), but not in the opposite direction.

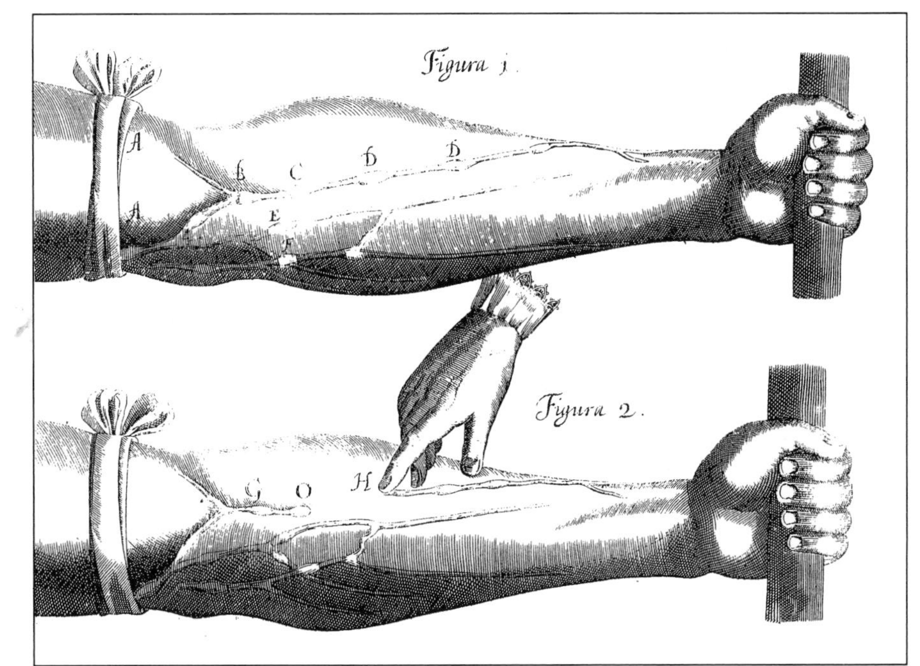

FIG. 13.5 *Diagrams from William Harvey's (1628) book* Exercitatio anatomica de motu cordis et sanguinis in animalibus *(An anatomical dissertation on the movement of the heart and blood in animals).*

The heart rate is most easily measured by feeling the pulse at the wrist and counting the number of beats in a minute. At rest, the heart beats about 45 times per minute in endurance athletes and rather faster (commonly 70 to 80 times per minute) in other adults. A low heart rate is sufficient in resting athletes because their large hearts deliver more blood in each stroke. In strenuous exercise it beats about 200 times per minute, both in athletes and in other reasonably healthy people (but it circulates blood faster in athletes because their hearts are bigger). The maximum pumping rates are about 40 litres per minute for endurance athletes and 20 litres per minute for sedentary young men. One method of measuring these rates is to take samples of arterial and venous blood and measure the oxygen concentrations in them. The difference between these concentrations multiplied by the rate at which blood is being pumped through the heart is the rate at which the body is using oxygen, so if the oxygen consumption has also been measured the pumping rate can be calculated.

Each litre of blood that leaves the lungs and travels through the left side of the heart carries about 0.16 to 0.20 litres of oxygen. Much of this diffuses out from the capillaries, as the blood that returns to the right atrium contains only about 0.12 litres of oxygen, (at rest) or as little as 0.02 litres (in strenuous activity) in each litre. A group of top athletes running as fast as they could for five minutes on a moving belt pumped blood through their hearts at an average rate of 36 litres per minute and used 5.6 litres of oxygen per minute. They were extracting 5.6/36 = 0.16 litres of oxygen from each litre of blood.

Blood leaves the heart in spurts, one for each heartbeat, so the pressure in the arteries fluctuates. Medical practitioners use sphygmomanometers (FIG. 13.6) to measure the range of blood pressures in an artery in the arm. They inflate the cuff until the pressure in it is enough to squash the artery flat and

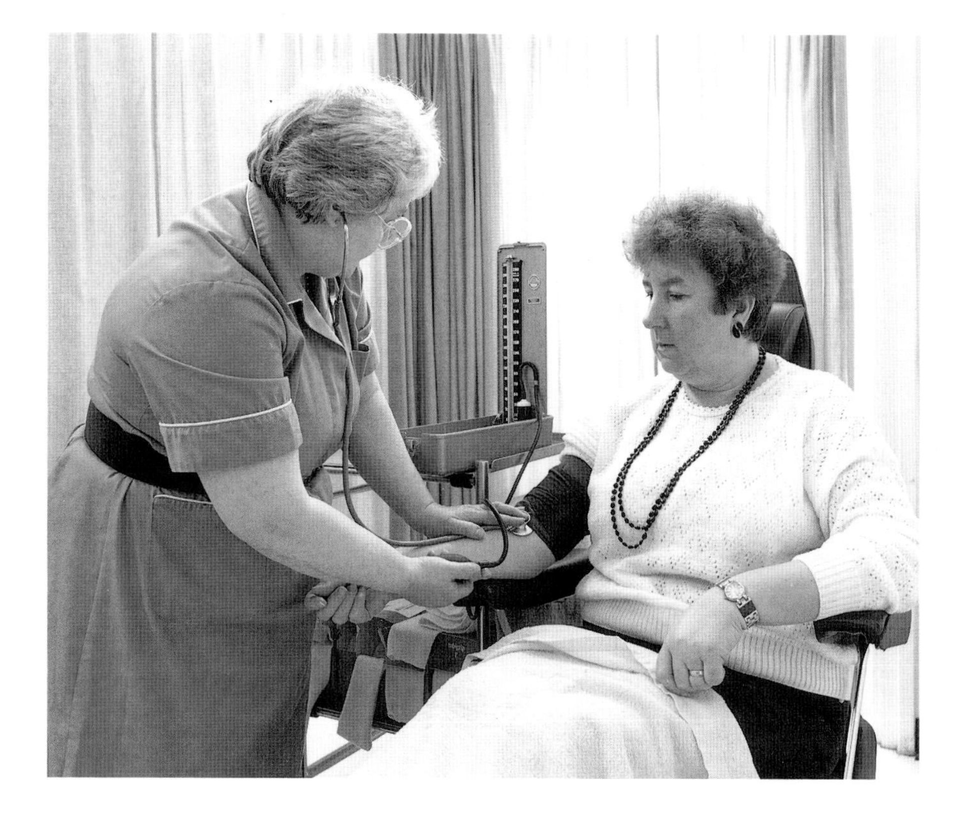

FIG. 13.6 *A sphygmomanometer in use.*

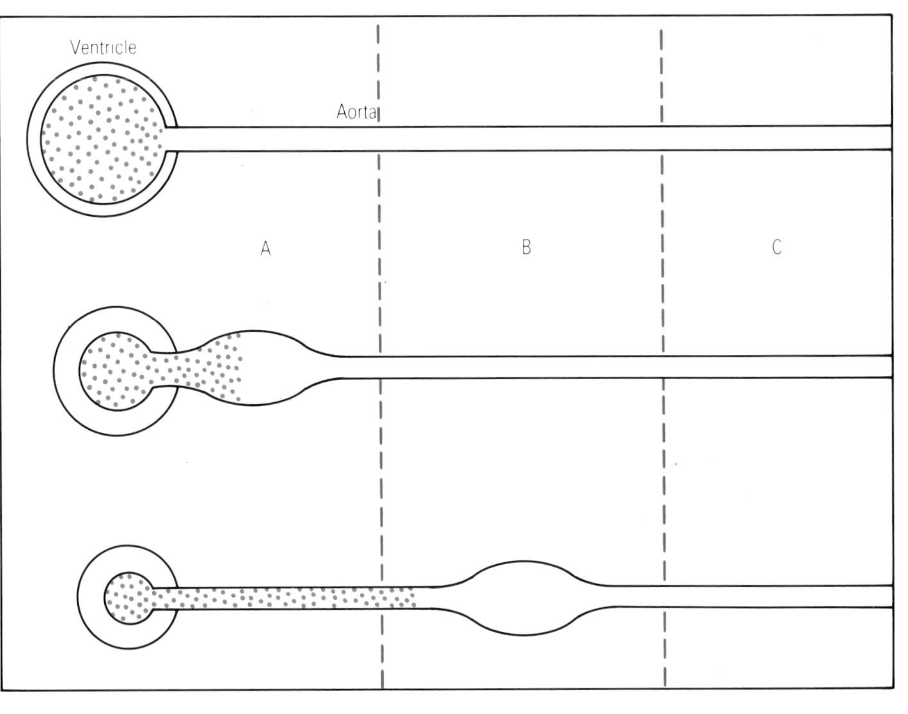

FIG. 13.7 *A diagram of the pulse wave.*

stop the flow of blood. They then gradually release the pressure while listening with a stethoscope. When the pressure in the cuff falls below the maximum blood pressure, the artery opens and blood flows for part of each heart beat cycle. It vibrates as it closes, and these vibrations are detected as sounds. When the pressure in the cuff falls below the minimum blood pressure the artery remains open all the time and the sounds disappear. The physician records the pressures in the cuff when the sounds appear and disappear: these are approximately the maximum and minimum arterial pressures and are generally about 16000 and 11000 newtons per square metre, in healthy patients. They are usually expressed in different units, as about 120 and 80 millimetres of mercury, because sphygmomanometers of traditional design use a column of mercury to measure the pressures.

Experiments on animals, in which tubes connected to pressure transducers are inserted into the heart, show that the pressure in the left ventricle rises to the same peak as in the arteries but falls to zero (that is, to atmospheric pressure) while the ventricle is filling. The arterial pressure remains high at this stage, while the valve between ventricle and aorta is closed, because the arteries are elastic: they swell while the ventricle is emptying into them, and the pressure in them falls only slowly as the blood escapes from them into the capillaries. In dogs the aorta (the big artery that leads out of the left ventricle) swells in diameter by about seven per cent, and contracts again, in each heart beat. Arteries owe their elastic properties to having a large proportion of elastin (a rubber-like protein) in their walls, as well as collagen and smooth muscle.

The maximum pressure in the right ventricle, and in the arteries that lead from it to the lungs, is much less than in the left ventricle, about 2000 newtons per square metre (15 millimetres of mercury). Blood must flow through both sides of the heart at the same rate, if it is not to accumulate in one place, but lower pressures are sufficient on the right side because the lung circulation offers less resistance to blood flow than the circulation through the rest of the body.

The arterial system does not all swell at once, when the left ventricle empties into it. Instead, a wave of swelling (the pulse wave) travels through it. It starts off at about five metres per second in the aorta and speeds up to about ten in the smaller arteries. This is far faster than the flow of the blood, which enters the aorta at one to one and a half metres per second.

Think of the aorta as consisting of a series of segments (A,B,C etc., FIG. 13.7). Contraction of the ventricle drives extra blood into segment A, making it swell. This segment recoils sending some of its blood on into segment B, which in turn swells; and so on. FIG. 13.7 shows how this process can make the pulse wave travel much faster than any of the blood. Propagation of the wave depends on interaction between the mass of the blood and the elastic properties of artery walls, much as sound is propagated through air by interaction of its mass and elasticity.

The muscular walls of the heart need a good blood supply to bring oxygen to them. It is not enough to have blood in the atria and ventricles: the muscle is permeated by blood capillaries, branches of the coronary arteries.

Many middle-aged people develop atheromas, which are deposits of fat (largely cholesterol esters) in the walls of major arteries. Indirectly, these cause about half the deaths in developed countries. They form swellings that project into the bore of the artery. The layer of cells that covers the inner surface of all arteries may break down over the swelling, stimulating a reaction in the blood very like the clotting that occurs when blood seeps from damaged skin. A clot-like lump of blood cells tangled in protein fibres (called a thrombus) is apt to form at the atheroma. While it stays there it generally does not do much harm but it may get washed away by the blood flow and stick in finer vessels, blocking them. A thrombus that gets stuck in the brain, temporarily cutting off blood flow from part of the brain, causes a stroke. One that gets stuck in the arteries of the heart starves part of the heart muscle of blood and may weaken or even stop the beating of that part within 15 seconds. It may cause the brief paroxysms of chest pain known as angina pectoris or the longer-lasting pain of a heart attack, and is often fatal.

Curiously, atheromas seem most apt to develop in places where the flow of blood is sluggish. At a fork in an artery, the blood's momentum carries most of it close to the inner walls of the branches leaving a patch of sluggish flow near each outer wall (FIG. 13.8a). At a curve, its momentum makes flow faster on the outside and more sluggish on the inside of the bend (FIG. 13.8b). These regions of more sluggish flow are where atheromas are most likely to develop. They are commoner in people who have high arterial blood pressure, who have a high concentration of cholesterol in the blood and who smoke cigarettes, than in others.

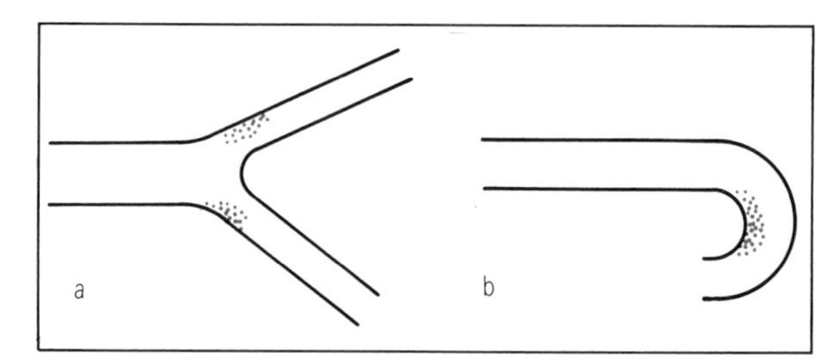

FIG. 13.8 *Blood flow is sluggish in the regions marked by stipple.*

When people are resting, with their hearts pumping about five litres of blood per minute, a large proportion of that blood goes to the brain and kidneys, typically about one litre per minute to each. A further litre per minute or thereabouts goes to each of two much larger masses of tissue, the digestive organs and the muscles. In heavy work the output of the heart increases to 20 litres per minute or more but the flows to brain, kidneys and digestive organs are little changed: nearly all of the huge increase goes to the muscles (where oxygen is needed to sustain activity) and to the skin (to get rid of excess heat produced by the rapid metabolism of the muscles). At rest, little blood flows through the skin, which serves as an insulating layer, helping to conserve heat. When we are overheated, plenty of blood flows through it and the insulation is bypassed. Flushing is the visible sign of the extra blood in the skin (FIG. 13.9).

In a house, you control the flow of water to different parts of your central heating system by adjusting valves which can be opened to allow free flow or shut down to restrict it. Flow in large parts of the system is controlled by these few small bottlenecks. Similarly, the distribution of flow in our blood systems is controlled by contracting or relaxing the muscle in the walls of arterioles (tiny arteries), constricting them or allowing them to dilate.

To see why the small arteries should be the bottlenecks that control flow we need to consider the lengths, diameters and numbers of blood vessels of each size. The pressure needed to drive blood through a blood vessel (that is, the pressure difference between its two ends) is proportional to:

$$\frac{\text{(vessel length)} \times \text{(rate of flow)}}{\text{(vessel diameter)}^4}$$

(Rate of flow means the volume passing through in unit time.) It seems obvious that the longer or the thinner the vessel, and the faster the flow, the more pressure will be needed, but the formula makes it clear that vessel diameter has a particularly strong effect.

The circulation can be thought of as a series of stages, one aorta opening into several large arteries which open into many small branches, and so on. It would be a huge task to work out the details for the whole circulation, but it is relatively easy to study the circulation in the mesentery, the transparent sheet of tissue that connects the gut to the body wall. TABLE 13.1 describes this circulation in a dog.

FIG. 13.9 *The difference of skin colour between the overheated athlete and the cool spectators is due to the control of blood flow through the skin. © Colorsport.*

	diameter d (mm)	number n	length l (mm)	l/nd^4	total area $\pi nd^2/4$ (cm^2)
large arteries	3	40	200	0.06	3
small arteries	1	2400	50	0.02	5
arterioles	0.02	4×10^7	2	0.31	125
capillaries	0.008	1.2×10^9	1	0.20	600
venules	0.03	8×10^7	2	0.03	570
small veins	2	2400	50	0.001	30
large veins	6	40	200	0.004	11

TABLE 13.1 *Numbers and dimensions of blood vessels in the mesentery of a dog. The data are from A.C. Burton (1972)* Physiology and Biophysics of the Circulation *ed. 2. Year Book, Chicago*

Obviously, it gives a extremely simplified view (the real thing is much less orderly, with stages merging into each other, and a lot of variation in length within each stage), but it will be good enough for a rough calculation. The total rate of flow must be the same for each stage but it is divided between different numbers of vessels, so the pressures needed to drive blood through successive stages are proportional to

$$\frac{\text{(vessel length)}}{\text{(number of vessels)} \times \text{(diameter)}^4}$$

The table shows this quantity for each stage, in the column labelled l/nd^4. It is largest for the arterioles and capillaries so if the arterioles constrict (making it substantially larger) blood flow in the mesentery will be greatly reduced. Direct measurements of pressure in the blood vessels of cat mesenteries have confirmed that a large proportion of the total pressure drop occurs in the arterioles and capillaries.

The cross-sectional area of a blood vessel of diameter d is $\pi d^2/4$ (π is 3.14). If a stage of the circulation consists of n vessels of this diameter, the total of their cross-sectional areas is $\pi nd^2/4$. TABLE 13.1 shows that this is much larger for the capillaries than for the arteries. The same blood flows through all stages of the circulation so the speed at which it travels (distance per unit time) is inversely proportional to $\pi nd^2/4$. Blood flows fast in arteries but very much slower in capillaries.

Oxygen is carried around the body in combination with the pigment haemoglobin, contained in red blood cells, which gives blood its red colour. These cells are by far the most plentiful in the blood, in terms of volume: they occupy about 45% of the blood volume (or less, in anaemic people), other cells occupy 1% and the rest is plasma, a solution of protein, salts etc. in water. The red cells are discs, relatively thick near the edge and thin in the middle. Their diameter is about eight micrometres which may seem too much, as they have to pass through capillaries of only five micrometres diameter. However, they are easily squashed out of shape. Blood can be seen flowing through capillaries when some parts of living animals are examined under the microscope: it can be seen, for example, in the ears of rabbits. Photographs show some of the red cells rolled up, and others distorted into bell shapes as in FIG. 13.10. Red cells pass undamaged through filters with pore diameters as small as three micrometres.

There seems to be a real advantage in having such tight-fitting cells. Imagine cell-free blood (with the haemoglobin dissolved in it) flowing in a capillary. It would flow smoothly, each haemoglobin molecule travelling more or less parallel to the capillary wall. (Fluids swirl around in irregular turbulent fashion when flowing fast through wide tubes, but flow smoothly when moving slowly through narrow ones.) Oxygen carried by haemoglobin molecules travelling down the centre of the capillary would have a relatively long way to diffuse to get to the capillary wall and so out to the tissues. To make matters worse, flow would be faster near the centre than near the walls (because viscosity would slow down flow near the walls) so a disproportionately large number of molecules would be kept well away from the walls.

Because the red cells fit tightly in the capillaries, only a little plasma leaks past them (but that little lubricates their passage). The plasma is fairly effectively divided into discrete chunks, separated by cells. Within each chunk, the plasma near the capillary wall is slowed by viscosity, so moves backward *relative to the cells* (FIG. 13.10) forcing plasma near the centre to move forward (again, relative to the cells). This circulation of plasma aids diffusion of oxygen between the red cell centres and the capillary wall. This effect seems likely to be important in the lungs, where oxygen has only to diffuse through 0.2 micrometres (1/5000 millimetres) of epithelium and capillary wall to reach the blood, and an extra few micrometres diffusion through the capillary would be a relatively large increase. It cannot be very important elsewhere because capillaries are generally at least 30 micrometres apart; much of most tissues is 10 micrometres or more from the nearest capillary, so diffusion distances within the capillaries are relatively trivial.

While the heart beats, it keeps the pressure in arteries far above that of surrounding tissues, and they remain inflated. The pressures in veins are much lower and may be greater or less than those of their surroundings. When the pressure inside is substantially greater than outside, a vein swells to a circular cross-section, but when it is substantially less the vein is squashed flat. Thus the volumes of veins vary far more than those of arteries.

To understand when veins are likely to swell and when to collapse, we have to think about hydrostatic pressure. The pressure increases as you dive down into water, by 10000 newtons per square metre (75 millimetres of mercury) for every metre increase of depth. Blood is only a little denser than water so there would be the same pressure difference between levels, in a stationary column of blood. To make blood flow upwards in a vein (for example, from a foot towards the heart) the pressure difference between the two ends of the vein must be greater than the hydrostatic pressure difference: it must be more than 75 millimetres of mercury, for every metre of height.

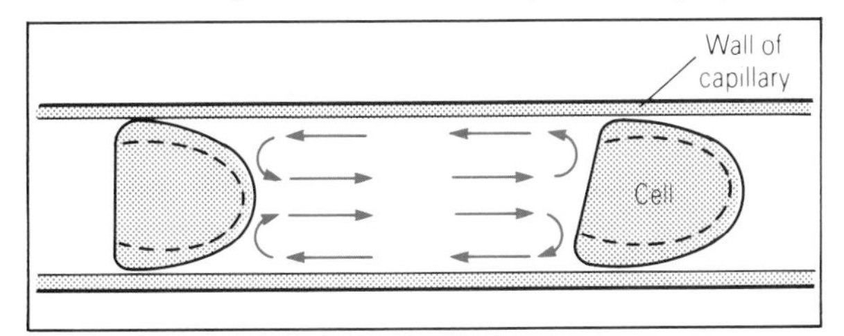

FIG. 13.10 *Red blood cells travelling from left to right along a capillary. The arrows show plasma flow* **relative to the cells.**

When someone is lying down all their blood vessels are fairly nearly at the same height but when they stand up, important differences of hydrostatic pressure appear. Compare blood vessels near the heart with others, one metre away, in a leg. While lying down, blood pressure in large arteries is about the same in both places (probably about 100 millimetres of mercury, averaged over a heart beat cycle, above atmospheric pressure) and blood pressure in veins in both places is only a little above atmospheric. When standing, the pressures may remain about the same near the heart but leg artery pressure will then rise to 175 millimetres of mercury because of the height difference, and leg vein pressure must rise above 75 millimetres of mercury. That may seem to present no problem because the pressure difference available to drive blood from arteries to veins remains unchanged, but it does have an important effect. Blood will not flow downwards in the leg veins because they have valves that prevent reverse flow, but it will not flow upwards in them until incoming blood from the capillaries has swollen them sufficiently to raise the pressure in them by 75 millimetres of mercury. Excessive swelling of leg veins is normally prevented by contraction of the muscle in their walls and also by activity in the leg muscles (i.e. in the striated muscles that work the leg skeleton). Contractions of leg muscles squeeze the veins, temporarily increasing the pressure in them enough to drive blood on past the next valve, so blood can be returned to the heart without the veins becoming swollen. When it occurs, however, excessive swelling can have striking effects. Aeroplane passengers who take off their shoes and rest with little leg movement are apt to find their feet so swollen, at the end of a long flight, that they have trouble getting their shoes on again. Soldiers on parade and others who stand still for long periods are apt to faint because accumulation of blood in their legs reduces flow to the brain.

CHAPTER FOURTEEN

SPEAKING

The human voice is a wind instrument that works like a trumpet but can produce much more subtle variations of sound. To understand it we need to know something of the physics of wind instruments. We also need some understanding of vibrating strings because a trumpeter's lips, and our vocal cords, vibrate rather like the strings of pianos and guitars.

FIG. 14.1 shows some of the ways in which a stretched string can vibrate: in each case the continuous line and the broken one are its position at the two extremes of the vibration. The fixed ends of the strings are nodes: that is, they remain stationary as the rest of the string vibrates. In the fundamental mode, they are the only nodes, and there is maximum amplitude of vibration (an antinode) half way along the string. In this mode, the frequency of vibration is lower than in any other. In the first harmonic there is an additional node, half way along the string, there are two antinodes and the frequency is twice the fundamental. In the second harmonic there are two intermediate nodes and the frequency is three times the fundamental. Other modes that I have not illustrated have four, five, six etc. times the fundamental frequency. A piano string does not generally vibrate in just one of these modes but makes a complex vibration that is a mix of the fundamental with many harmonics. The frequency that we identify as the pitch of the string, however, is the fundamental. The tighter the string, the higher this frequency is. (You tune pianos and guitars by tightening or loosening the string.) Also the longer or the thicker the string (for the same tension) the lower its fundamental frequency.

A noise is an irregular pattern of vibration, transmitted through the air, but a musical sound is a pattern that repeats itself regularly at least for a few cycles of vibration. Any musical note can be broken down into a component at the fundamental frequency and harmonics at 2, 3 etc. times that frequency. There is a general tendency for high-numbered harmonics of the sound emitted by a vibrating string to be less loud than low-numbered harmonics, but the decline of loudness with increased harmonic number is not steady.

Fundamental and harmonic vibrations also occur in wind instruments, but there may not be harmonics at every whole-number multiple of the fundamental. FIG. 14.2a shows air vibrating in an organ pipe that is closed at one end but open at the other. It vibrates freely in and out of the open end, so there is an antinode there, but there is a node at the closed end where the end wall prevents it from vibrating. This is the only node in the fundamental mode of vibration, but there are additional nodes in harmonic modes. There are harmonics only at odd-numbered multiples of the fundamental frequency: three (FIG. 14.2c), five (14.2d) etc. Notice that the fundamental in FIG. 14.2 has just one segment of vibration (from a node to an antinode) and that successive harmonics have three, five etc. Wide and narrow tubes of equal length have almost the same fundamental frequency but longer tubes have lower fundamentals than short ones: if you double the length of the tube you halve the frequency. The frequency is also affected by any bulges or constrictions in the tube.

The vocal cords have the same function in voice as the lips have in playing the trumpet: we blow air between them to set them vibrating. They are in the larynx or Adam's apple, a swelling in the trachea (FIG. 14.3). The cricoid cartilage is a complete ring

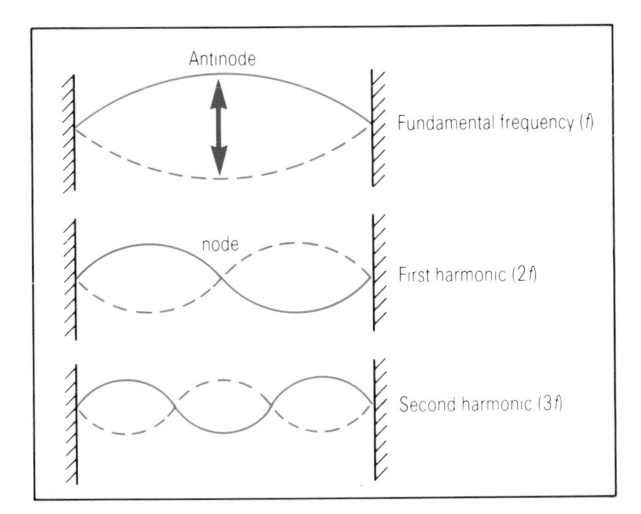

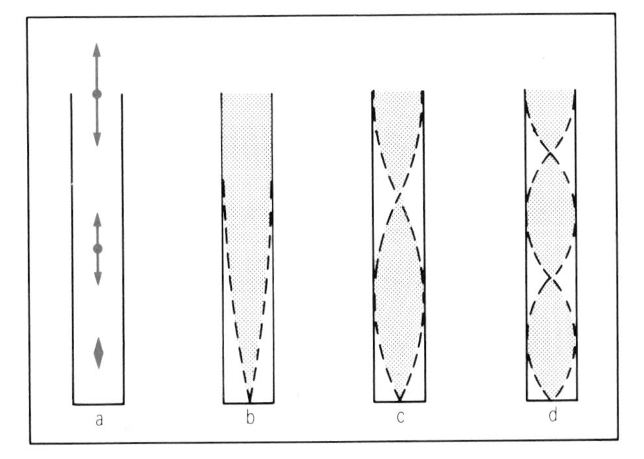

but the thyroid cartilage is an incomplete one, open at the rear. The two small arytenoid cartilages lie in the thyroid cartilage's gap, connected by movable joints to the cricoid cartilage. There are also movable joints between the cricoid cartilage and the lower horns of the thyroid. The vocal cords are membranous folds that protrude like shelves across the larynx, running from the thyroid cartilage to the arytenoids.

The two sets of cricoarytenoid muscles (shown in FIG. 14.3) rotate the arytenoid cartilages away from each other to separate the vocal cords (FIG. 14.4c) or towards each other to make them meet and close the larynx. Contraction of the cricothyroid muscles rocks the cricoid cartilage on the thyroid in such a way as to tighten the vocal cords (FIG. 14.4), and the thyroarytenoid muscles slacken them.

It is possible to look down the larynx to see the vocal cords in action. The simplest method is to put a mirror into the subject's mouth, but that makes it difficult for him or her to speak normally. A more sophisticated method uses a slender, flexible fibre-optic bundle that will transmit images around bends. If it is pushed up the speaker's nose and through to the back of the mouth that way, it interferes very little with speech. By looking through

FIG. 14.1 *(top) Vibrations of a stretched string.*

FIG. 14.2 *(above) Vibrations of air in a tube. The amplitude of the vibrations in different parts of the tube is represented in (a) by the lengths of the arrows and in (b), (c) and (d) by the width of the tinted band.*

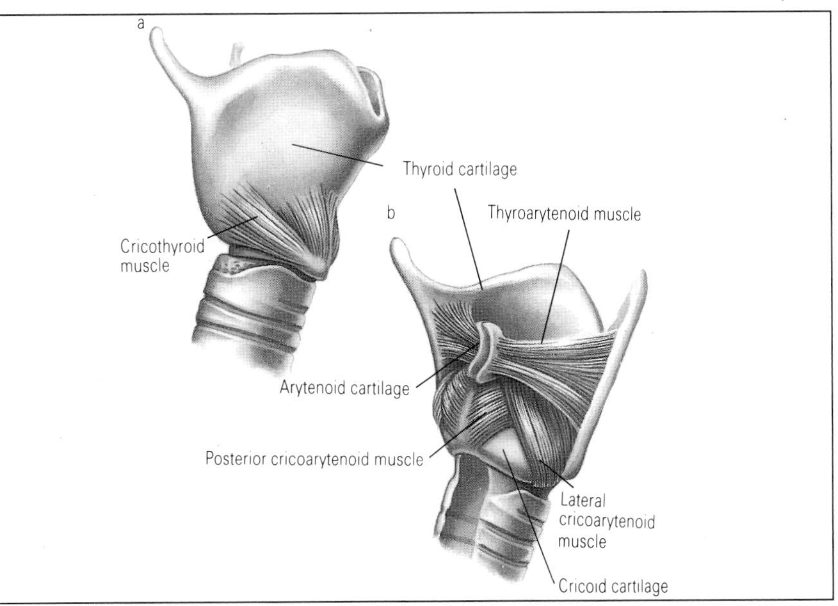

FIG. 14.3 *Cartilages and muscles of the larynx.*

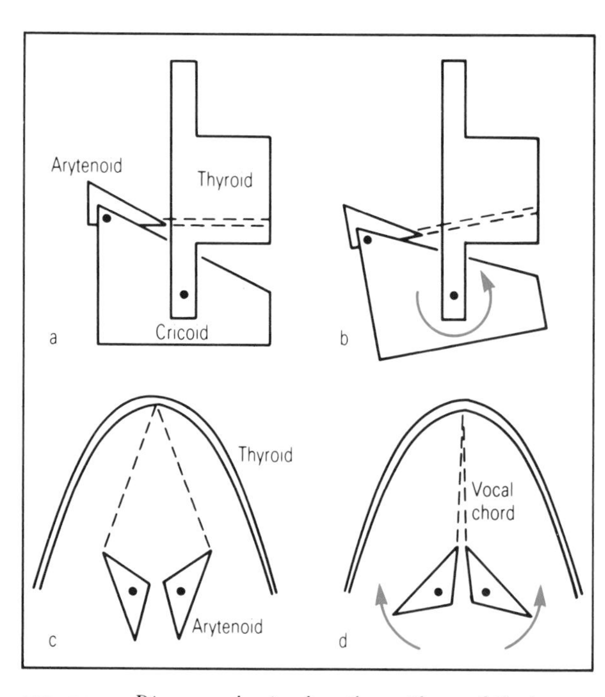

FIG. 14.4 *Diagrams showing how the cartilages of the larynx can be moved. (a) and (b) are side views, and (c) and (d) are seen from above.*

such devices it can be seen that when people are not talking, their vocal cords are well separated, leaving a wide passage for breathing air in and out. During speech they move closer together, leaving only a narrow slit and at times closing the larynx completely (FIG. 14.5). High speed cine films show them vibrating in complicated ways.

Johannes Müller was an outstanding pioneer in experimental physiology, who published a textbook in several volumes in the 1830s. In it he describes experiments that he had performed on larynges taken from human carcases. He hung weights from them to simulate the pull of the muscles and blew air through. When the vocal cords were tight and closed to a narrow slit, sounds were produced which rose in pitch if the tension was increased. Just as tightening a guitar string raises its pitch, tightening the vocal cords raises theirs, moving the fundamental and all the harmonics to higher frequencies. The average fundamental frequency of ordinary speech is about

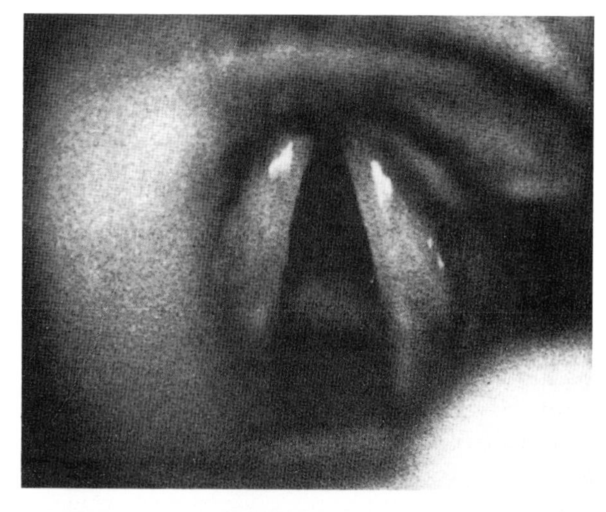

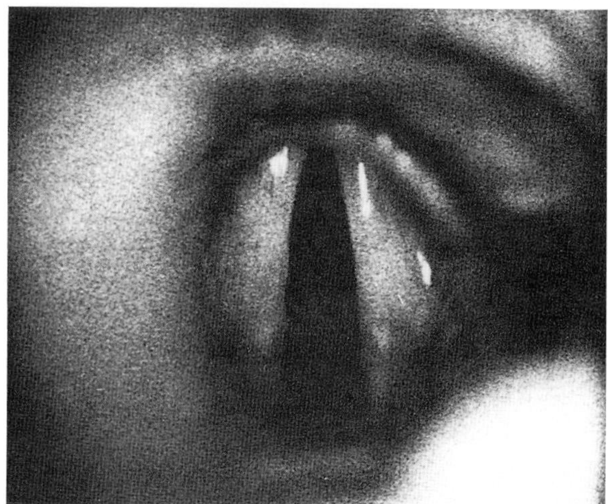

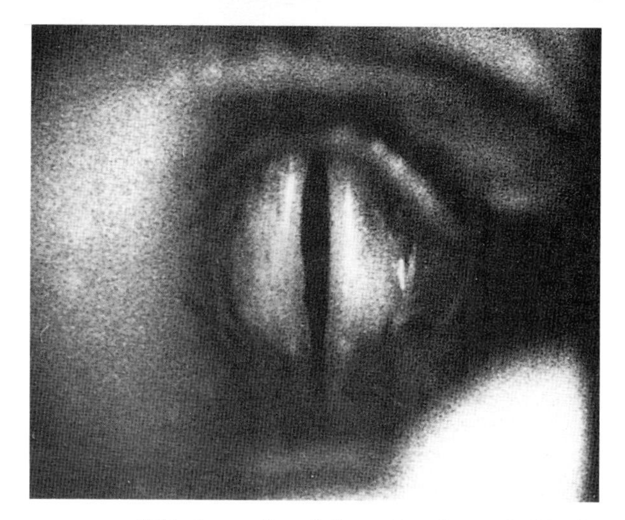

FIG. 14.5 *Stills from a film of the vocal cords, supplied by Dr P. Lieberman.*

90 cycles per second for bass voices and 280 for sopranos. The entire singing range in Mozart's operas is from 73 to 1400 cycles per second.

Adult men have deeper voices than women and children because their larynges grow rapidly at puberty. The vocal cords get longer and therefore vibrate at a lower fundamental frequency, at any particular tension. This physical difference makes it difficult for most men to produce a passable imitation of a woman's voice.

Until after the time of Shakespeare, women's parts in plays were played by young men. That has puzzled some people who have felt that young boys with unbroken voices would be too immature (intellectually and emotionally) to perform well in intense dramatic roles such as Cleopatra or Lady Macbeth, and that hulking young men with broken voices and a five o'clock shadow would be as unconvincing as pantomime dames. However my colleague Dr Richard Rastall has pointed out that though modern boys' voices usually break at the age of 13 or 14 years they used to break later. Written records of Bach's choir in 1744 show that the boys' voices used to break then at 16 to 18. A century and a half before that, in Shakespeare's time, many eighteen year old boys may have sounded convincingly feminine. They may also have looked feminine because the breaking voice, growth of the beard and the other changes that go with them are all effects of puberty.

The air must speed up as it passes through the narrow gap between the vocal cords. Where it travels faster its pressure is less (this is called the Bernoulli effect) so it tends to pull the vocal cords together, making the gap even narrower. The lift on aeroplane wings is another consequence of the Bernoulli effect: the wings are shaped and angled so as to speed up the air over their upper surfaces, so the pressure is higher below the wing than above it. As the Bernoulli effect narrows the slit between the vocal cords the air flow is eventually slowed and the pressure rises again. The vocal cords spring apart, the air flow speeds up again and the cycle is repeated. Thus the air flow drives the vibration of the vocal cords and their vibration makes the air flow intermittent. The air comes through them as a series of exceedingly short puffs and each puff is a sound wave.

As this implies, we speak only while breathing out. A cycle of normal quiet breathing lasts about four seconds, just under two seconds breathing in and just over two breathing out. While speaking we make many of our outward breaths much longer, in extreme cases up to forty seconds. We breathe in between sentences, or where commas would appear if what we are saying were written.

The vibrations that are transmitted through air as sound have energy, and the intensity of sound is defined as the power per unit area: for example, the intensity of sound at your open window is the sound energy coming through it in unit time, divided by the area of the window. It is sometimes given as you might expect in watts per square metre but more often in rather confusing units called decibels. Zero on the decibel scale is usually defined as one millionth of a millionth of a watt per square metre: this tiny quantity is approximately the threshold of human hearing, the intensity of a sound that is just audible. Every decibel means an increase of intensity by a factor of 1.26, so 10 decibels is $(1.26)^{10} = 10$ times the threshold intensity, 20 decibels is $(1.26)^{20} = 100$ times the threshold, 30 decibels is $(1.26)^{30} = 1000$ times the threshold and so on. Typical sound intensities are 30 decibels in a quiet room and 100 decibels inside a rock band. One advantage of the decibel scale is that it corresponds well with our subjective impressions of loudness: 100 decibels seems about as much louder than 90 decibels, as 90 seems louder than 80.

People who study speech are generally not content to know the overall sound intensity: they want to know the frequencies of the fundamental and of all the harmonics, and the intensity of each. One type of analyser examines the sound during a short time interval and produces a graph of average intensity during the interval, against frequency (FIG. 14.6). Another produces a record showing frequency plotted against time, with darker marks indicating higher intensities (FIG. 14.7).

FIG. 14.8 shows idealized sound spectra with the peaks represented by lines. The sound from an

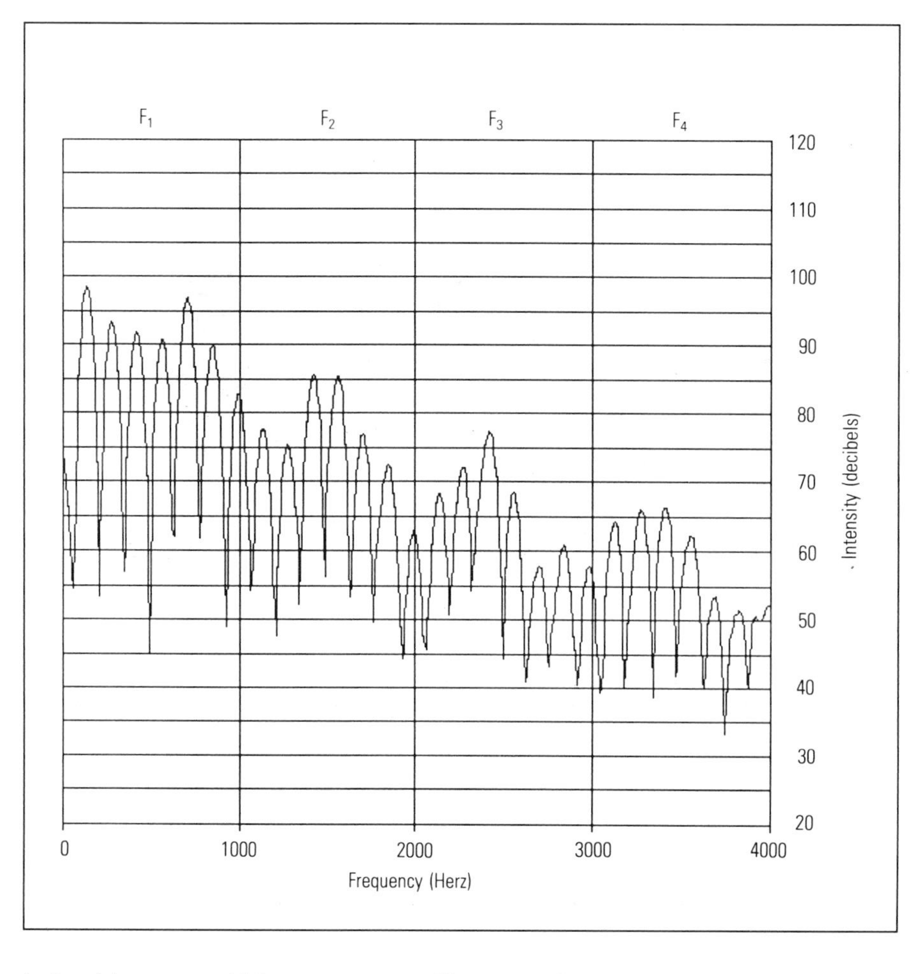

FIG. 14.6 *A sound spectrum of the vowel sound 'a' from 'pattern' spoken by a British man. F1 to F4 mark the formants. This record was supplied by Dr Peter J. Roach and Ms Celia Scully of the Department of Linguistics and Phonetics, University of Leeds.*

isolated larynx would have a spectrum like FIG. 14.8a: the fundamental would have the highest intensity and the harmonics progressively lower ones. Actual spectra are more complicated because our larynges are connected to the pharynx, mouth and nasal cavities, whose resonances enhance some frequencies and suppress others. Suppose that a larynx which by itself produced the spectrum shown in (a) were connected to a uniform tube, 165 millimetres long, open at the other end. (You will soon see why I choose this particular length.) This tube would have a fundamental frequency of 500 cycles per second and harmonics at three, five, seven etc. times this frequency (FIG. 14.2). When the larynx emitted sound into it, the tube would enhance its harmonics at 500, 1500, 2500 etc. cycles per second and suppress intermediate harmonics (FIG. 14.8b). If the vocal cords were then tightened to increase their fundamental frequency to 200 cycles per second, their harmonics would be spaced twice as far apart but those near 500, 1500 etc. cycles per second would still be enhanced (FIG. 14.8c).

The spectra in FIG. 14.8b, c are very like those of people voicing the vowel sound in 'heard'. To make this sound, we adjust the mouth opening and the position of the tongue so as to make a tube of fairly uniform cross-section, from larynx to lips. The length of this tube in an adult is about 165 millimetres (a little more in men, a little less in women), making its fundamental frequency about 500 cycles

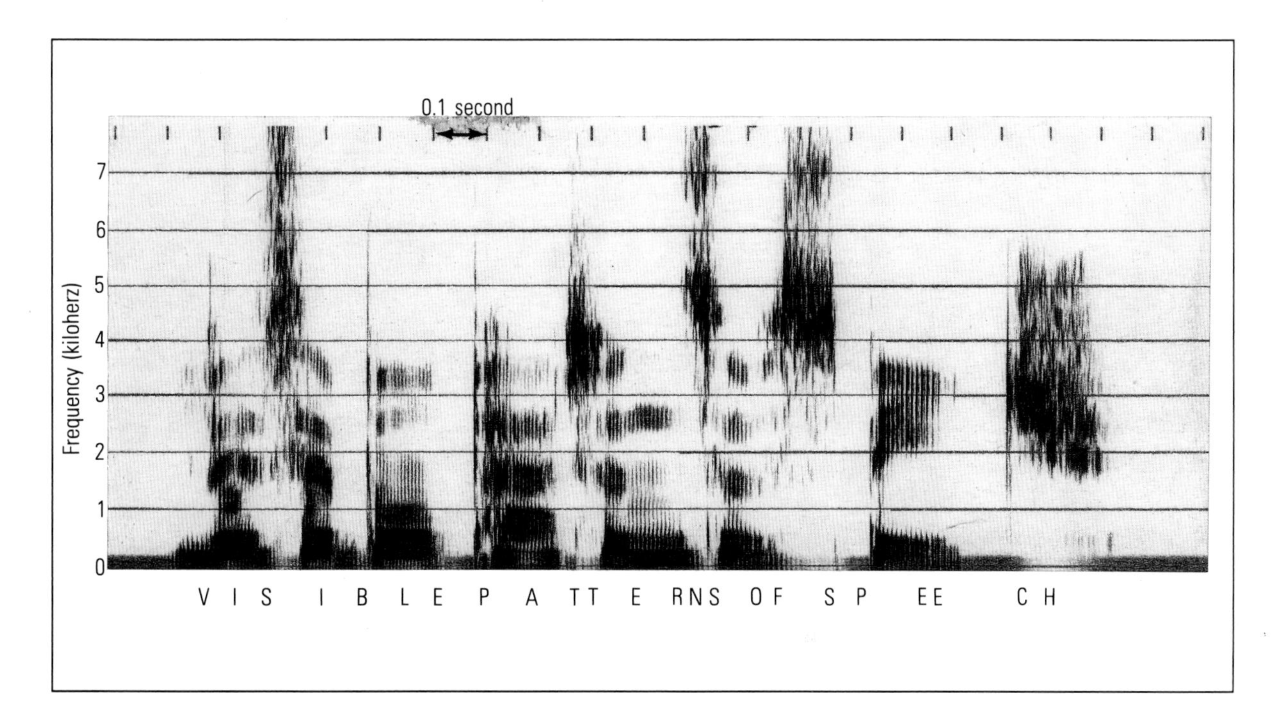

FIG. 14.7 *A sound spectrogram of the phrase 'visible patterns of speech', spoken by a British man. This record was supplied by Ms Celia Scully and Dr Peter J. Roach of the Department of Linguistics and Phonetics, University of Leeds.*

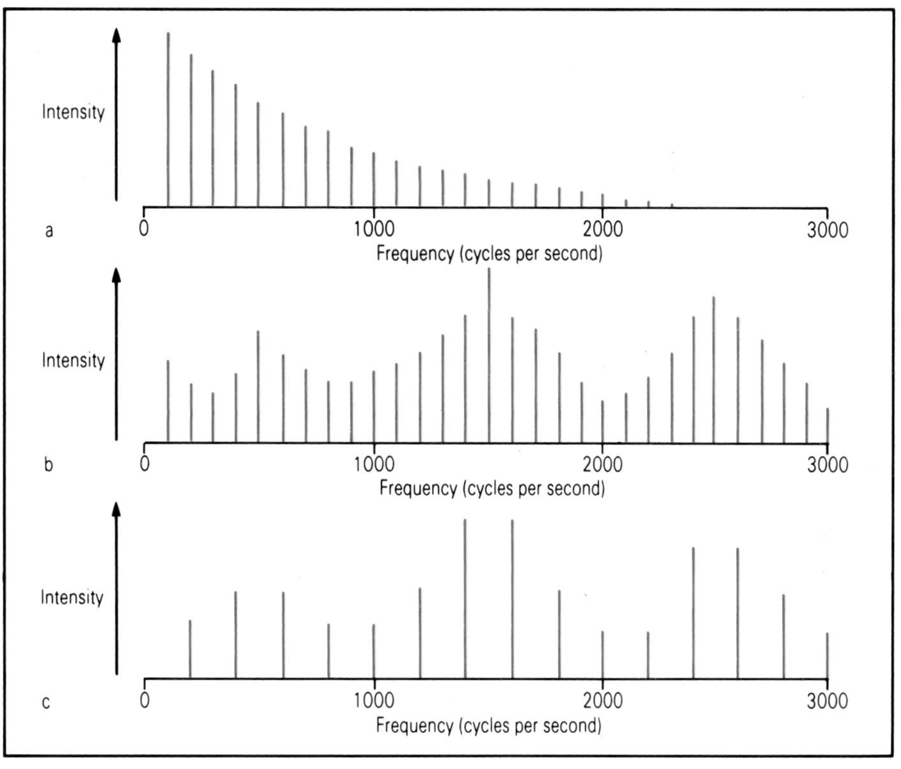

FIG. 14.8 *Schematic sound spectra. The fundamental frequency is 100 cycles per second (a) and (b), 200 in (c). In (b) and (c) there are formants at 500, 1500 and 2500 cycles per second.*

per second. The frequencies it enhances (500, 1500 etc. cycles per second) are called the formants. When we sing the same vowel at different pitches we change the fundamental frequency of the vocal cords but keep the formant frequencies unchanged, as in FIG. 14.8b, c. People of different sizes, however, have different formant frequencies. A five year old child who measured only 110 millimetres from larynx to lips instead of the 165 that we have been assuming would produce a first formant of $500 \times 165/110 = 750$ cycles per second, by the movements that in an adult would produce 500.

To make other vowels we put our mouths and tongues in other positions, so that the passage from larynx to lips is not a uniform tube, and enhances different formant frequencies. FIG. 14.9 shows some mouth positions. Tongue positions have been investigated by taking X-ray pictures of people with small pieces of metal stuck to their tongues. To produce the vowel o of 'hot' you open your mouth to enlarge the mouth cavity but move the base of your tongue back to narrow the pharynx. That makes the path from larynx to lips start narrow and widen out, rather as in FIG. 14.10a. This diagram shows two

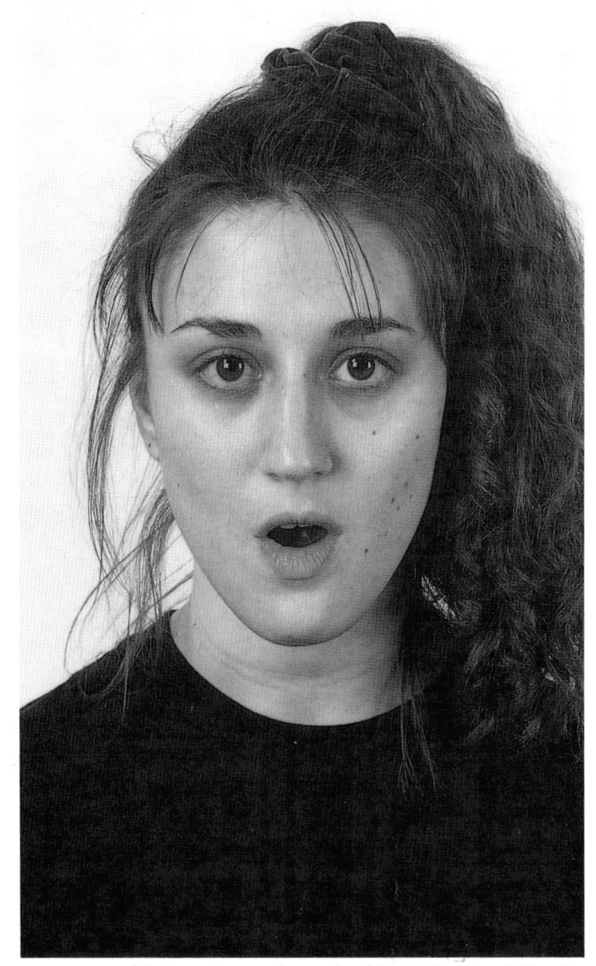

FIG. 14.9 *Making the vowel sounds of 'hot', 'heat' and 'hoot'.*

tubes in series. The narrow 'pharynx' is effectively open at the mouth end because the other tube is so much wider, and the wide 'mouth cavity' is effectively closed at the pharynx end because the pharynx is so much narrower. The combination resonates at frequencies close to the fundamental frequencies that the tubes would have if they were separate as in FIG. 14.10b, c, but not exactly the same as if they were separate. Even if the two tubes were equal in length, the combination would give pairs of formant frequencies, one 1.4 times the other. The path from larynx to lips has narrow and wide parts

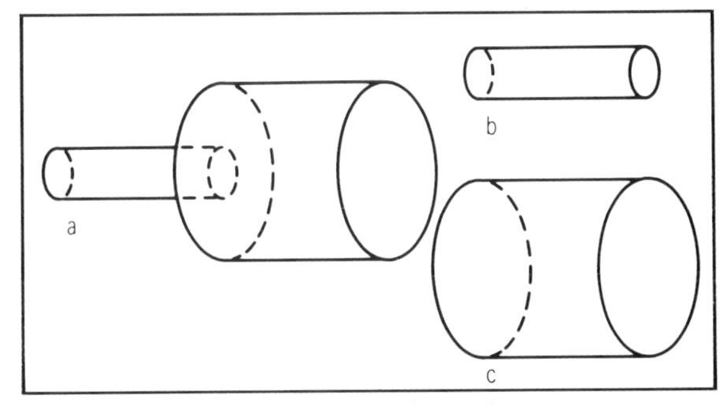

FIG. 14.10 *Two tubes joined together, and separate.*

of about equal length, when we pronounce the o of 'hot', giving formants close together, at about 800 and 1100 cycles per second. Notice how much closer these are than the first two formants of the vowel of 'heard', at 500 and 1500 cycles per second.

FIG. 14.11 shows the frequencies of the first two formants for American English vowels. 76 people were given lists of the words shown on the graph, in random order, and asked to read them out loud. They were recorded and their vowel sounds were

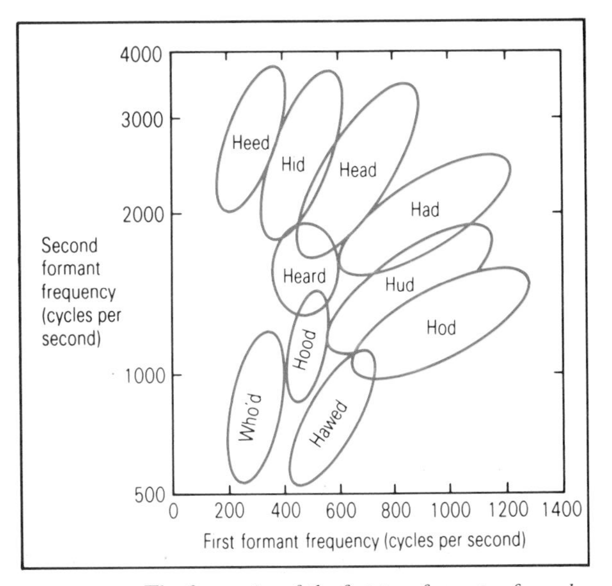

FIG. 14.11 *The frequencies of the first two formants of vowel sounds in (American) English. Each loop encloses about 90% of the values recorded when 76 people each read out the words. From G.E. Peterson & H.L. Barney (1952)* J. acoust. Soc. Am. ***24,*** *175–184.*

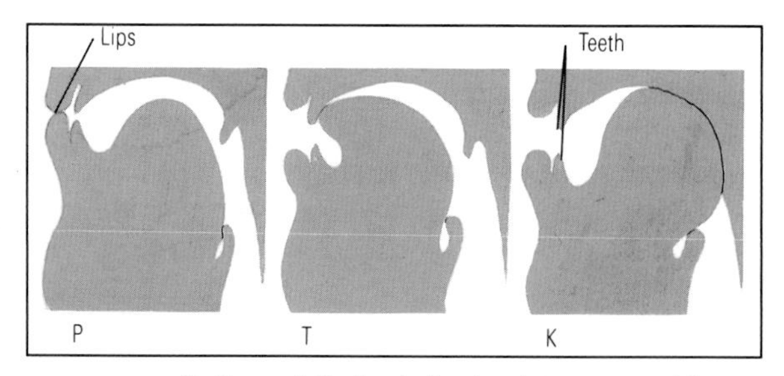

FIG. 14.12 *Sections of the head showing how* p, t, *and* k *are spoken. From P. Leiberman & S.E. Blumstein (1988)* Speech Physiology, Speech Perception and Acoustic Phonetics *Cambridge University Press, Cambridge.*

analysed. There was variation between one reader and another and even some overlap (for example, 'heard' pronounced by one speaker may be the same as 'hood' pronounced by another), but the graph shows that vowels can generally be distinguished by their first two formant frequencies. Different languages use different sets of vowel sounds, with different formant frequencies. Some of the French vowels are notoriously difficult for English speakers.

Consonants are produced in various ways. The plosive consonants are made by blocking the mouth cavity, letting pressure build up, and then releasing it suddenly. For *k* you use the back of the tongue to block the rear part of the mouth, for *t* the blockage is made by the tongue on the teeth and for *p* it is made by the lips (FIG. 14.12). *g*, *d* and *b*, respectively, are made by releasing the same three blockages, but the vocal cords move together sooner so that their vibrations start sooner after the release. These plosive consonants produce rapidly-changing sound frequencies that merge with the following vowel.

Synthetic sounds have been used to discover the essential character of these consonants. FIG. 14.13

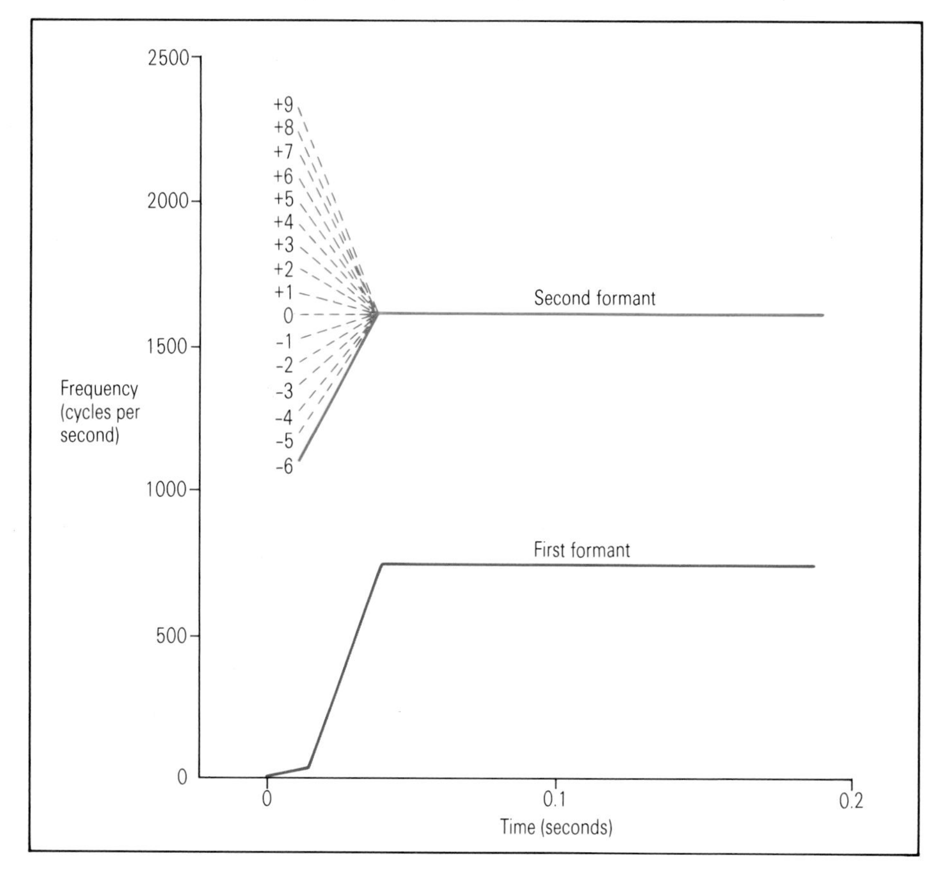

FIG. 14.13 *A diagram showing the patterns of sound (frequency plotted against time) used in an investigation of the voiced plosive consonants* g, d *and* b.

shows how the sounds 'ga', 'da' and 'ba' (as in 'gad', 'dad' and 'bad') were imitated. In each case the vowel was represented by its first two formants, at 750 and 1600 cycles per second. One component of the sound always rose rapidly from an initial low frequency to merge with the first formant of the vowel, but the second formant started either with a fall in frequency (alternatives $+9$ to $+1$) or with a rise (-1 to -6). Listeners identified sounds $+6$ to $+9$ as 'ga', 0 to $+4$ as 'da' and -2 to -6 as 'ba'. Intermediate sounds were ambiguous, identified sometimes as one of the possibilities and sometimes the other.

Another group of consonants, the fricatives, involve hissing sounds made by forcing air through small gaps. *f* uses gaps made by setting the upper incisor teeth against the lower lip, and *s* gaps made by putting the tip of the tongue behind the upper incisors. *sh* is made by blowing through a gap between the tongue and the roof of the mouth and *h* by blowing through the vocal cords while they are held wider apart than for vowel sounds but not as wide as for normal breathing. The characteristic of the fricatives is noise (random vibration, rather than regularly repeating patterns) due to turbulence in the air as it escapes from the constriction. *v* and *z* are voiced versions of *f* and *s*.

The nasals are consonants that use the resonance of the nasal passages. The mouth cavity is blocked at some point but the velum, the flap at the back of the palate, is opened to let air escape through the nose. The passage from larynx to nostril is longer than from larynx to lips and it is bulbous, wide inside but constricted at the nostrils. It gives a low-pitched first formant, at about 250 cycles per second. For *m* the mouth is closed at the lips, for *n* it is closed by the tongue pressing against the roof of the mouth and for *ng* (as in 'thing') further back. There is a low murmur due to nasal resonance which ends as the block is opened.

Opera singers make themselves heard over the orchestra, with no need for the microphones carried by singers of less exalted music. How do they manage this? Spectral analysis shows that the sounds they produce have a strong formant at a frequency of about 2500 cycles per second, a frequency at which the sound of an orchestra is rather weak. This is apparently produced by lowering the larynx and widening the pharynx.

Coughing is described here rather than in the chapter on breathing, because it involves the vocal cords. It produces a brief, very rapid flow of air to clear unwanted material from the air passages in the lungs. We cough to clear out food that has got into the air passages accidentally, or to drive out excess mucus produced by inflamed tissues. A cough starts with an inward breath. The vocal cords are then closed and pressure is built up in the thorax and abdomen, by the muscles that are normally used for breathing out. This pressure may rise to 13,000 newtons per square metre or even more, similar to the pressures developed in the abdomen in weight-lifting and in forceful defaecation (and also to arterial blood pressure). The cough occurs when this pressure is released suddenly by opening the vocal cords.

A cough may drive out a litre of air, in one tenth of a second. Flow at this rate through the trachea, which has a cross-sectional area of two square centimetres, implies that the speed of the air is 50 metres per second. It must reach considerably higher speeds in air passages that collapse for a reason that I will explain. High speeds are needed to lift mucus from the walls of the air passages so that it can be carried away as a mist-like suspension of tiny droplets.

Air passages in the lungs are apt to collapse during a cough, because of the Bernoulli effect (which was explained earlier in this chapter). If a slight constriction forms in one of these tubes while air is flowing through, the air must flow faster and so (by the Bernoulli effect) the pressure must fall. If it falls too low the tube will collapse, stopping the flow temporarily. This process may make the walls of the tube vibrate, and help to shake off mucus. It also sets an upper limit to the speed at which air can be made to flow through. It is impossible to drive air through a collapsible tube faster than the speed at which waves (like pulse waves in arteries) travel along it.

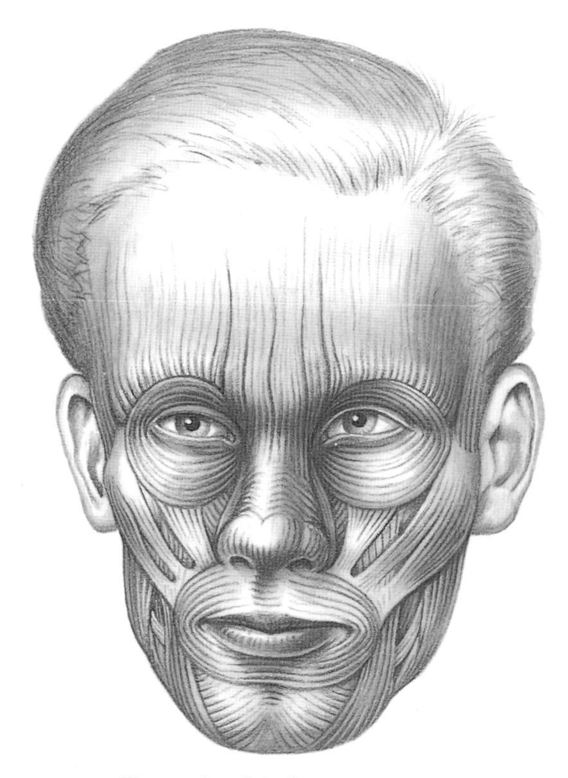

FIG. 15.1 *The muscles of the face.*

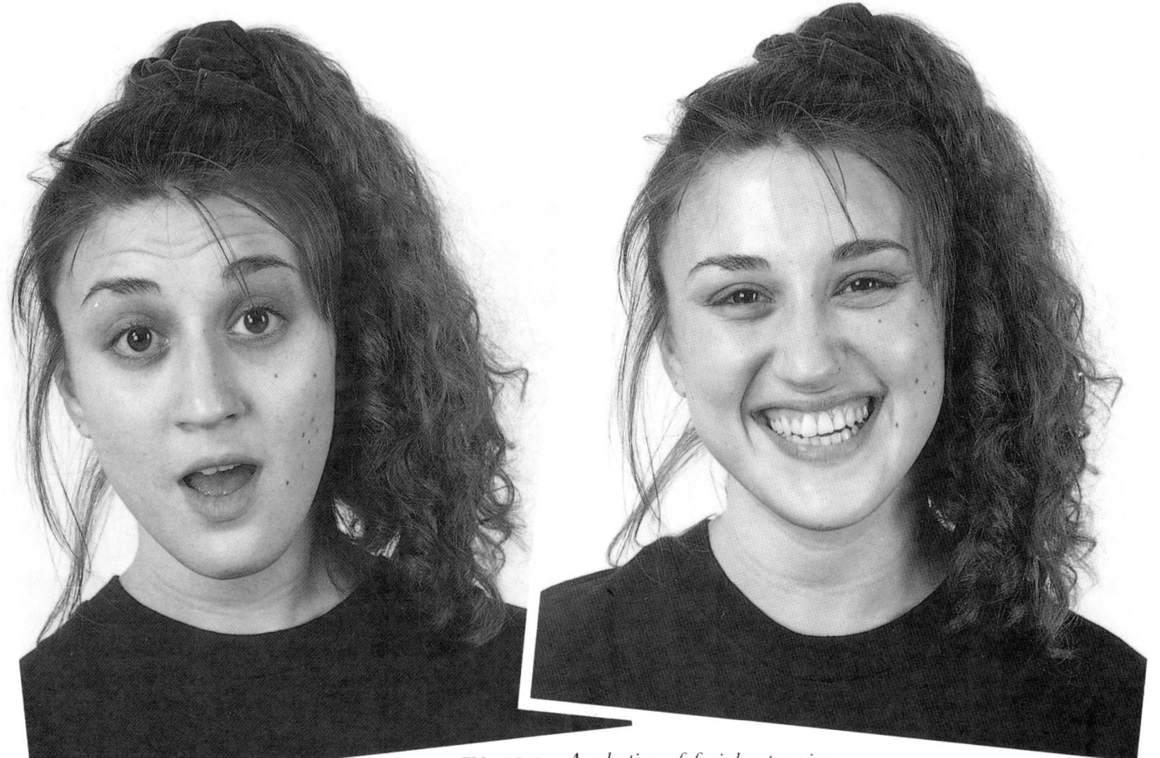

FIG. 15.2 *A selection of facial expressions.*

CHAPTER FIFTEEN

THE HUMAN MACHINE

This book has ranged from weightlifting to operatic singing, but cannot deal with all aspects of the working of the human body. It has shown that mechanics, as developed by engineers, is a valuable aid to understanding our bodies but has ignored other important branches of physical science. We need optics to explain our eyes, electrical theory to understand nerves and the theory of heat exchange to discuss the temperatures of our bodies. Chemistry is needed to explain the processes of digestion and the exceedingly complicated metabolic processes that occur inside our cells. All these have been ignored in this book, which is concerned solely with mechanics.

Even its coverage of mechanics is incomplete. I have concentrated on movements that we can perform at will, paying little attention to those that we cannot control consciously such as the heart beat, the movements of our guts and the beating of the microscopic cilia on the walls of our respiratory passages. I have also omitted many interesting applications of mechanics to sense organs, including the organs of balance and of hearing in our ears, and the muscle spindles that tell us what our muscles are doing.

Even some interesting types of voluntary movement have been omitted. Here is just one, to end the book. FIG. 15.1 shows the muscles of the face (these are striated muscles, under voluntary control) and FIG. 15.2 shows some facial expressions. Can you work out for yourself which muscles must contract to produce each expression?

PRINCIPAL SOURCES

Here is a list of many, but by no means all, of the books and papers consulted in the course of writing this book. Information from each of those in the GENERAL section was used in several chapters.

GENERAL

Alexander, R.McN. (1983) *Animal Mechanics* ed. 2. Blackwell, Oxford.

Åstrand, P.-O. and K. Rodahl (1986) *Textbook of Work Physiology* ed. 3. McGraw-Hill, New York.

Ballreich, R. and A. Kuhlow (1986) *Biomechanik der Leichtathletik* Enke, Stuttgart.

Basmajian, J.V. and C.J. de Luca (1985) *Muscles Alive.* Williams and Wilkins, Baltimore.

Dickson, R.A. and V. Wright (1984) *Musculoskeletal Disease* Heinemann, London.

McFarlan, D. (editor) (1989) *The Guiness Book of Records 1990* Guiness Publishing, London.

McMahon, T.A. (1984) *Muscles, Reflexes and Locomotion.* Princeton University Press.

Nordin, M. and V.H. Frankel (1989) *Basic Biomechanics of the Musculoskeletal System.* Lea and Febiger, Philadelphia.

Williams, P.L. and R. Warwick (editors) (1980) *Gray's Anatomy* ed. 36. Churchill Livingstone, Edinburgh.

CHAPTER 1

Pheasant, S. (1986) *Bodyspace: Anthropometry, Ergonomics and Design.* Taylor and Francis, London.

CHAPTER 2

Hadler, N.M. *et al.* (1978) Hand structure and function in an industrial setting. *Arthritis & Rheumatism 21*: 210–220.

Mason, M.T. and J.K. Salisbury (1985) *Robot Hands and the Mechanics of Manipulation.* M.I.T. Press, Cambridge, Mass.

Napier, J.R. (1980) *Hands.* Allen & Unwin, London.

Unsworth, A.D. Dowson and V. Wright (1971). Cracking joints. *Annals of the Rheumatic Diseases 30*, 348–358.

CHAPTER 3

Hutton, W.C. and M.A. Adams (1982) Can the lumbar spine be crushed in heavy lifting? *Spine 7*, 586–590.

Ker, R.F., R.McN. Alexander and M.B. Bennett (1988) Why are mammalian tendons so thick? *Journal of Zoology 216*, 309–324.

Poppen, N.K. and P.S. Walker (1978) Forces at the glenohumeral joint in abduction. *Clinical Orthopaedics 135*: 165–170.

Schultz, A. *et al.* (1982) Loads on the lumbar spine. *Journal of Bone and Joint Surgery 64a*: 713–720.

Troup, J.D.G. and F.C. Edwards (1985) *Manual of Handling and Lifting.* H.M.S.O. London.

CHAPTER 4

Debrunner, H.U. (1985) *Biomechanik des Fusses* Enke, Stuttgart.

Ellis, M.I., B.B. Seedhom, V. Wright and D. Dowson (1980) An evaluation of the ratio of the tensions along the quadriceps tendon and the patellar ligament. *Engineering in Medicine 9*; 189–194.

Heglund, N.C. and G.A. Cavagna (1987) Mechanical work, oxygen consumption and efficiency in isolated frog and rat striated muscle. *American Journal of Physiology 253*: C22–C29.

Markolf, K.L., J.S. Mensch and H.C. Amstutz

(1976) Stiffness and laxity of the knee. *Journal of Bone and Joint Surgery* **58a**: 583–594.

McLeish, R.D. and J. Charnley (1970) Abduction forces in the one-legged stance. *Journal of Biomechanics* **3**: 191–209.

Molleson, T. (1989) Seed preparation in the Mesolithic: the osteological evidence. *Antiquity* **63**: 356–362.

O'Connor, J., J. Goodfellow and E. Biden (1981) Designing the human knee. In I.A.F. Stokes (editor) *Mechanical Factors and the Skeleton* 52–64. Libbey, London.

Thomasen, E. (1982) *Diseases and Injuries of Ballet Dancers* Århus University Press.

CHAPTER 5

Alexander, R.McN. (1984) Walking and running. *American Scientist* **72**: 348–354.

Alexander, R.McN. (1991) Characteristics and advantages of human bipedalism. In J.M.V. Rayner & R.J. Wootton (editors) *Biomechanics in Evolution.* In press. Cambridge University Press.

Bornstein, M.N. and H.G. Bornstein (1976) The pace of life. *Nature* **259**: 557–558.

Clarke, M. (1975) Materials problems in high-heeled shoes, and the influence of style and gait. *Journal of the British Boot and Shoe Institution* Nov.–Dec. 1975, 171–178.

Hay, R.L. and M.C. Leakey (1982) The fossil footprints of Laetoli. *Scientific American* **246(2)**: 38–45.

Jackson, K.M. (1983) Why the upper limbs move during human walking. *Journal of Theoretical Biology* **105**: 311–315.

Mochon, S. and T.A. McMahon (1980) Ballistic walking: an improved model. *Mathematical Biosciences* **52**: 241–260.

Waters, R.L. and B.R. Lunsford (1985) Energy cost of paraplegic locomotion. *Journal of Bone and Joint Surgery* **67a**: 1245–1250.

CHAPTER 6

Alexander, R.McN. (1988) *Elastic Mechanisms in Animal Movement.* Cambridge University Press.

Alexander, R.McN. and M.B. Bennett (1989) How elastic is a running shoe? *New Scientist* **123** (1673): 45–46.

Cavagna, G.A., L. Komarek and S. Mazzoleni (1971) The mechanics of sprint running. *Journal of Physiology* **217**: 709–721.

Cavanagh, P.R. (editor) (1990) *The Biomechanics of Distance Running* Human Kinetics Publishers, Champaign.

Cavanagh, P.R. and M.A. Lafortune (1980) Ground reaction forces in distance running. *Journal of Biomechanics* **13**: 397–406.

Greene, P.R. (1987) Sprinting with banked turns. *Journal of Biomechanics* **20**: 667–680.

McMahon, T.A. and P.R. Greene (1978) Fast running tracks. *Scientific American* **239(6)**: 148–163.

Nigg, B.M. (1986) *Biomechanics of Running Shoes* Human Kinetics Publishers, Champaign.

Nigg, B.M., H.A. Bahlsen, S.M. Luethi and S. Stokes (1987) The influence of running velocity and midsole hardness on external impact forces in heel-toe running *Journal of Biomechanics* **20**: 951–959.

Pugh, L.G.C.E. (1971) The influence of wind resistance in running and walking and the mechanical efficiency of work against horizontal or vertical forces. *Journal of Physiology* **213**: 255–276.

Wieser, W. and E. Gnaiger (editors) (1989) *Energy Transformations in Cells and Organisms* Thieme, Stuttgart.

CHAPTER 7

Alexander, R.McN. (1990) Optimum take-off techniques for long and high jumpers. *Philosophical Transactions of the Royal Society* B**329**: 3–10.

Bobbert, M.F. and G.J. van Ingen Schenau (1988) Coordination in vertical jumping. *Journal of Biomechanics* **21**: 249–262.

Cavagna, G.A., B. Dusman and R. Margaria (1968) Positive work done by a previously stretched muscle. *Journal of Applied Physiology* **24**: 21–32.

Dapena, J. and C.S. Chung (1988) Vertical and radial motions of the body during the take-off phase of high jumping. *Medicine and Science in Sports and Medicine* **20**: 290–302.

Frolich, C. (1980) The physics of twisting somersaults. *Scientific American* **242**: 112–120.

Hay, J.G. (1986) The biomechanics of the long jump. *Exercise and Sport Science Reviews* **14**: 401–446.

CHAPTER 8

Davis, R.R. and M.L. Hull (1981) Measurement of pedal loading in bicycling. II. Analysis and results. *Journal of Biomechanics* **14**: 857–872.

de Prampero, P.E. (1986) The energy cost of human locomotion on land and in water. *International Journal of Sports Medicine* **7**: 55–72.

Pugh, L.G.C.E. (1974) The relation of oxygen uptake and speed in competition cycling and comparative observations on the bicycle ergometer. *Journal of Physiology* **241**: 795–808.

Stuart, M.K., E.T. Howley, L.B. Gladden and R.H. Cox (1981) Efficiency of trained subjects differing in maximal oxygen uptake and type of training. *Journal of Applied Physiology* **50**: 444–449.

Vandewalle, H. *et al.* (1987) Force-velocity relationship and maximal power on a bicycle ergometer. *European Journal of Applied Physiology* **56**: 650–656.

Whitt, F.R. and D.G. Wilson (1982) *Bicycling Science* ed. 2. M.I.T. Press, Cambridge, Mass.

CHAPTER 9

Adrian, M.J., M. Singh and P.V. Karpovich (1966) Energy cost of leg kick, arm stroke, and whole crawl stroke. *Journal of Applied Physiology* **21**: 1763–1766.

Holmér, I. (1979) Physiology of swimming man. *Exercise and Sport Science Reviews* **7**: 87–123.

Martin, R.B., R.A. Yeater and M.K. White (1981) A simple analytical model for the crawl stroke. *Journal of Biomechanics* **14**: 539–548.

Miller, D.I. (1975) Biomechanics of swimming. *Exercise and Sport Science Reviews* **3**: 219–248.

Pendergast, D.R. *et al.* (1977) Quantitative analysis of the front crawl in men and women. *Journal of Applied Physiology* **43**: 475–479.

CHAPTER 10

Alexander, R.McN. (1991) Optimum timing of muscle activation for simple models of throwing. *Journal of Theoretical Biology* **150**: 349–372.

Atwater, A.E. (1979) Biomechanics of overarm throwing. *Exercise and Sport Science Reviews* **7**: 43–85.

Bartlett, R.M. and R.J. Best (1988) The biomechanics of javelin throwing: a review. *Journal of Sports Science* **6**: 1–38.

Daish, C.B. (1972) *The Physics of Ball Games* English Universities Press, London.

Elliott, B.C. and D.H. Foster (1984) A biomechanical analysis of the front-on and side-on fast bowling techniques. *Journal of Human Movement Studies* **10**: 83–94.

Zatsiorsky, V.M., G.E. Lanka and A.A. Shalmanov (1981) Biomechanical analysis of shot putting technique. *Exercise and Sport Science Reviews* **9**: 353–389.

Zernicke, R.F. and E.M. Roberts (1978) Lower extremity forces and torques during systematic variation of non-weight bearing motion. *Medicine and Science in Sports* **10**: 21–26.

CHAPTER 11

Davenport, H.W. (1982) *Physiology of the Digestive Tract* ed. 5. Year Book Medical Publishers, Chicago.

Gibbons, C.P., E.A. Trowbridge, J.J. Bannister and N.W. Read (1988) The mechanics of the anal sphincter complex. *Journal of Biomechanics* **21**: 601–604.

Shaw, J.H., E.A. Sweeney, C.C. Cappuccino and S.M. Meller (editors) (1978) *Textbook of Oral Biology* Saunders, Philadelphia.

Smith, K.K. and W.M. Kier (1989) Trunks, tongues and tentacles: moving with skeletons of muscle. *American Scientist* **77**: 29–35.

Weijs, W.A. (1990) The functional significance of morphological variation of the human mandible and masticatory muscles. *Acta Morphologica Neerlando – Scandinavica* **27**: 149–162.

CHAPTER 12

Braun, N.M.T., N.S. Arora and D.F. Rochester (1982) Force-length relationship of the normal human diaphragm. *Journal of Applied Physiology* **53**: 405–412.

Hills, B.A. (1988) *The Biology of Surfactant* Cambridge University Press.

Macklem, P.T. and J. Mead (editors) (1986) *Handbook of Physiology section 3, Volume III Mechanics of Breathing* American Physiological Society, Bethesda.

CHAPTER 13

Alpert, J.S. (1984) *Physiopathology of the Cardiovascular System* Little Brown, Boston.

Caro, C.G., T.J. Pedley, R.C. Schroter and W.A. Seed (1978) *The Mechanics of the Circulation* Oxford University Press.

Winkel, J. and K. Jørgensen (1986) Evaluation of foot-swelling and lower-limb temperatures in relation to leg activity during long-term seated office work. *Ergonomics* **29**: 313–328.

CHAPTER 14

Ainsworth, W.A. (1976) *Mechanism of Speech Recognition* Pergamon, Oxford.

Leiberman, P. and S.E. Blumstein (1988) *Speech Physiology, Speech Perception, and Acoustic Phonetics.* Cambridge University Press.

Rastall, R. (1985) Female roles in all-male casts. *Mediaeval English Theatre* **7**: 25–50.

Sundberg, J. (1977) The acoustics of the singing voice. *Scientific American* **236 (3)**, 82–91.

INDEX